Organization and Environment

Managing Differentiation and Integration

HARVARD BUSINESS SCHOOL CLASSICS

1. ORGANIZATION AND ENVIRONMENT: *Managing Differentiation and Integration*
   Paul R. Lawrence and Jay W. Lorsch

2. STRATEGY FOR FINANCIAL MOBILITY
   Gordon Donaldson

3. MANAGING THE RESOURCE ALLOCATION PROCESS: *A Study of Corporate Planning and Investment*
   Joseph L. Bower

4. DIRECTORS: *Myth and Reality*
   Myles L. Mace

5. STRATEGY, STRUCTURE, AND ECONOMIC PERFORMANCE
   Richard P. Rumelt

Organization and Environment

Managing Differentiation and Integration

PAUL R. LAWRENCE

and

JAY W. LORSCH

Both of the Harvard Business School

With the research assistance of

JAMES S. GARRISON

HARVARD BUSINESS SCHOOL PRESS

BOSTON, MASSACHUSETTS

The paper used in this publication meets the requirements of the American National Standard for Permanence of Paper for Printed Library Materials Z39.48–1984.

Harvard Business School Press, Boston 02163

 Revised edition
Printed in the United States of America

95 94 93 92 6 5 4

Library of Congress Cataloging-in-Publication Data

Lawrence, Paul R.
Organization and environment.

(Harvard Business School classics; 1)
Includes index.
1. Organizational effectiveness—Case studies.
2. Organizational behavior—Case studies. 3. Management—Case studies. I. Lorsch, Jay William. II. Title.
III. Series.
HD58.9.L39 1986 658.4'02 85–30490
ISBN O-87584-129-5

CONTENTS

PREFACE TO THE HARVARD BUSINESS SCHOOL CLASSICS EDITION

To us this book has become an interesting intellectual phenomenon. At the time it was published it won both the Academy of Management's Award for "best management book of the year," and the James A. Hamilton-Hospital Administrators' Book Award. In the twenty years since the study was performed, the book has been widely cited and is currently listed among the Best Management Books.* In short, it has become a "classic" worthy of being kept in print. Its stature has led us to consider two basic questions: why is the book noteworthy and what is its relevance in 1986 and beyond? Obviously our answers have a certain bias, but they represent our best assessment of the impact of the work after two decades.

The book's greatest significance is that it signaled a major shift in the paradigm in organizational behavior. Prior to 1967, the dominant question in the field had been what is *the single best* way to manage and organize? By introducing contingency theory, *Organization and Environment* changed the basic issue to "what management style and organization form is best suited to a particular situation?" More specifically, the focus was on the fit between an organization and its environment.

While we were the first to be recognized for pointing broadly to this new direction, the idea of situational thinking can be traced back as far as Fritz Roethlisberger and

* *Good Book Guide for Business, The Economist* (New York: Harper & Row, 1984).

William Dickson's *Management and the Worker* (1939) and Chester Barnard's *Functions of the Executive* (1938) in which the idea of an organization as a social system was introduced. And, as so often happens in intellectual history, a number of other scholars arrived at the same point we did in the mid-1960s: Tom Burns and G. M. Stalker, Joan Woodward, Lawrence Fouraker, Alfred Chandler, Stanley Udy, Harold Leavitt, Fred Fiedler, Victor Vroom, and Arthur Turner are mentioned in Chapter VIII. Of these, Burns and Stalker had the earliest and most profound effect on our work. It was from Fred Fiedler that we borrowed the term "contingency." Not mentioned but also pointing in the same direction at the time were James Thompson and James March and Herbert Simon. In our view then, *Organization and Environment's* "classic" status is a tribute not just to the study itself, but also to the fact that it was published at a time when all of these scholars were pointing to this basic shift in the organizational behavior paradigm.

Beyond this, we believe that the specific concepts of differentiation and integration have proven to be very powerful constructs; such validation of our theory has contributed to the book's durability. Scholars have used our ideas to understand hospitals, school systems, municipal governments, multinational and diversified companies, as well as the relationship between employees and their organizations. Managers and consultants concerned with organization design have also found these concepts to be very useful for analysis and action planning.

Another strength of the study is that it is truly multidisciplinary, focusing on phenomena of concern to both psychologists and sociologists. In fact, evidence of the parochial views of these disciplines can be found in the way scholars in each reacted to the book. Sociologists saw its virtues in the focus on matters structural, but ignored the attention paid to interpersonal issues, such as conflict resolution. Psychologists had the opposite reaction.

Mentioning such reactions suggests we would be remiss if we did not recognize that the book has also had its critics. Most criticisms centered on the fact that we seemed to be offering a model of the organization-environment relationships that was static and deterministic. We were interpreted as concluding that environmental characteristics determined the organizational form that would lead to effective results. And it was said that we gave no attention to the specific process by which changes in the environment were accommodated by changes in the organization. Similarly, scholars concerned with business policy and general management worried that our model had no room for the concept of strategy.

In one sense these criticisms are valid. We did focus the book on the way the environment impacts on the organization. Our use of the terms *environmental demands* and *requirements* clearly reveals this. Because of our view that organizations are systemic in nature (see pp. 6–7) we assumed the reader would understand that we recognized that choices made by managers inside the organization could impact on the environment. However, we were not explicit on this point. Also, because our research evidence provided one-time snapshots of the organization, it did have a static quality. But, our theory was dynamic and later we, and others, did address the relevant change process. Similarly, we now can see that we should have recognized more explicitly the role of strategic choice in determining what specific environment the organization encountered. We took corporate strategies as "given" only to limit the scope of our inquiry. We certainly were not in any sense "anti-strategy." Clear evidence of this is provided by our attention to Chandler's work as a contributor to contingency theory (pp. 195–198).

Other criticism centered on our method of measuring environmental characteristics. Unfortunately, this alleged shortcoming seemed to reveal more about the need of academics for research topics that could be suitably captured in short journal articles than it did about improving the means of assessing environmental characteristics.

We mention these criticisms not just to provide a balanced picture of the book but also because they suggest another reason to introduce this new edition. It is our hope that it will stimulate additional research into the organization-environment interface, so that the model can be further tested and the relationships more fully understood. What are better ways to understand and measure environmental characteristics? How do organizational forms inhibit or encourage adaption to environmental change? Where does the concept of strategy fit into these processes?

We also hope that this edition will renew interest in contingency theory. The general notion of fit has become almost axiomatic, as illustrated by McKinsey and Company's Seven S's popularized by *In Search of Excellence*. Yet much remains to be learned about designing "the discrete management practices" referred to in Chapter IX (pp. 224–228). The broad contingency approach, as useful and powerful as it is, needs refinement and precision. It is our hope that the continued availability of the book will aid such thinking and research. At the same time being pragmatic, we hope the book will stimulate practicing managers and consultants to continue using the specific concepts articulated as well as the general contingency model. This too should build the knowledge base that is needed to manage effectively our economic institutions.

Finally, in introducing this new edition we want to thank again all of those who contributed to our research—the managers who willingly agreed to be studied, our secretaries and editors, and the Dean and Directors of the Division of Research at the Harvard Business School who allowed us the time to conduct the study.

Paul R. Lawrence
Jay W. Lorsch

Boston, Massachusetts
January 1986

LIST OF TABLES

LIST OF FIGURES

CHAPTER I

Background and Approaches to the Study

MANAGERS have long recognized that different industrial environments have particular economic and technical characteristics, each of which calls for a unique competitive strategy. A set of marketing, manufacturing, and research policies that works well for a firm in the chemical industry will not meet the needs of a corporation producing steel. As obvious as these statements appear, their implications for organization theory have for too long been ignored. In this book we will make a connection between the varying technical and economic conditions outside the organization and the patterns of organization and administration that lead to successful economic performance. We will be seeking an answer to the fundamental question, *What kind of organization does it take to deal with various economic and market conditions?* We think this approach is both unexplored and important.

THE INADEQUACY OF CURRENT ORGANIZATION THEORY

During the past few years there has been no shortage of books and articles aimed at interpreting various aspects of man's behavior in organizations. Often, however, the research findings and ideas presented in one piece of literature have been difficult to relate to those in others. By implication or explicit recommendation the current literature suggests to the manager the utility of divergent managerial styles, organization structures and climates, and types of management training. For the behavioral and social scientists who devote their lives to understanding these topics, the apparent contradictions and ambiguities are confusing enough, but for the practicing administrator who is supposed to use this new knowledge as a guide in making organizational decisions, the confusion may at times seem insurmountable.

This confusion about organization theory is not surprising if we recognize that because large organizations are so complex, researchers usually study only segments and limited aspects of their operations. Few efforts have been made, until very recently, to understand their functioning as a whole. Most of the current findings and ideas have been developed from studies in different parts of large organizations. Some studies are centered in factories, while others look at research laboratories, but these studies often imply that the findings from one situation hold for all others. Similarly, those few studies that have examined total structures have been conducted in organizations performing a limited variety of tasks and facing a limited range of economic and technical conditions. Nevertheless, the findings of these studies have often been generalized to all organizations. The difficulty is that the essential organizational requirements for effective performance of one task under one set of economic and technical conditions may not be the same as those for other tasks with different circumstances. The organizational requirements for an effective sales unit may be quite different from those for an effective production unit. A firm producing a standard commodity sold to a few customers in a stable market may require a form and style of organization altogether unlike those of a company producing a highly sophisticated technical product for a more dynamic market.

In essence, much of the current organizational literature is directed at a fundamental question quite different from our own. Instead of seeking relationships between organizational states and processes and external environmental demands, as we are doing, most organizational research and theory has implicitly, if not explicitly, focused on *the one best way to organize in all situations.* We believe that seeking an answer to our question—What kind of organization does it take to deal with different environmental conditions?—may help to bring some order out of the current confusion about organization theory.

The need for such clarification is urgent. The problems discussed in the current literature go to the heart of immediate pragmatic organizational decisions. Managers must find answers to such questions as: What type of organization will best coordinate our sales effort? How much control and direction should we give our research scientists? Can improvements in our organization help us to develop more new products? What can we do to achieve better coordination between sales and plant personnel on delivery schedules? Will changes in our financial reward or control systems improve the effectiveness of our managers? These are the organizational issues about which managers are constantly deciding. Too often in the past these decisions have been made with no systematic analysis, or on the basis of generalizations about "the best way" to organize, or simply in imitation of a competitor who has adopted a particular organizational practice. The increasing knowledge about the functioning of organizations can provide useful tools for making sounder organizational decisions if some order can be brought to the present confused state of organization theory.

The need for reducing this theoretical confusion is vital for one more, equally important reason. One of the major causes of management's concern with organizational issues is that the technical, economic, and geographical conditions facing their organizations are becoming more diverse and

are constantly changing. The pace of technological change is stepping up, and the technologies of process and product are becoming more complex. Increasingly, corporations are becoming multinational, operating over wider geographical areas and under more diverse economic and cultural conditions. These trends have just started, and projections indicate that they will accelerate. For managers they raise a set of problems similar to those cited above, but perhaps even more complicated. What are the organizational implications of increasing rates of technological change? Should two product divisions dealing with different technologies and different markets be organized in the same way or differently? If differently, what should the difference be? Should an overseas subsidiary be organized like the U.S. parent?

We do not suggest that this book can answer all of these questions. It does, however, describe a research study that attempts to find out what types of organization will be effective under different economic and technical conditions. In so doing it offers a way of understanding the complexities of large organizations which can be helpful in making more sense out of some of the current organization theory and in providing a sounder basis for explicit managerial decisions about organizational matters.

The research findings and ideas presented here are addressed primarily to the practicing administrator, because they appear to apply immediately to his thinking about issues of formal organizational structures and procedures. We should, however, state clearly at the outset that the findings of this study seem to have important theoretical implications and suggest a number of opportunities for future inquiry. We hope, therefore, that the study reported here will also be of interest to other behavioral and social scientists and will stimulate them to further investigation of the functioning of large organizations.

The Major Concepts

The Choice of Concepts

Concepts have a common use for the manager and for the social scientist interested in the study of organizations. They are tools to examine and understand organizational phenomena. If concepts are to be useful to either the manager or the researcher, they must, it seems to us, be relatively few, with relatively simple relationships that can be understood. To quote a twentieth-century version of Occam's razor, "As few (concepts) as you may; as many (concepts) as you must." [1] For us as researchers it has seemed important to be economical in using concepts, because from a methodological standpoint we are only able to deal with and understand the relationship among a few variables at once. Similarly, this conceptual frugality is important for the manager confronted with decisions about organizations. Given the nature of man's thinking equipment, the variables must be few if they are to be mentally manipulated simultaneously. This happy coincidence of our needs as researchers and the needs of the practitioner is one reason that the findings of this study should be useful to managers.

Not all researchers have adopted this same strategy. Some have attempted to delineate all the possible variables in large organizations, without regard to the potential operational problems of measuring and relating them as researchers, or of being able mentally to juggle them as practitioners.[2] In the long run this strategy may pay off, but we have placed our bets on a different approach—that of using as few concepts as possible to find an answer to our fundamental question.

We should emphasize that we are dealing with a complex topic. We are not concerned, as others have been, only with understanding the already complicated subjects of the be-

[1] Source footnotes will be found at the back of the book.

havior of an individual manager or of a group of executives, but we are concerned with the functioning of large organizations, consisting of large numbers of individuals and many groups. This topic is even more complex, because our interest goes beyond the internal functioning of the organization to the more intricate problem of how what happens inside the organization is related to market and technical conditions outside the firm. The very complexity of the matters with which we are dealing has required that we use a number of concepts, but we have selected them with the principle of parsimony in mind. If the findings of this study are at times involved, it is because the territory they describe is exceedingly complex, and not because we have chosen to draw the most detailed map possible.

The Organization as a System

At the most general level we find it useful to view an organization as an open system in which the behaviors of members are themselves interrelated. The behaviors of members of an organization are also interdependent with the formal organization, the tasks to be accomplished, the personalities of other individuals, and the unwritten rules about appropriate behavior for a member. Under this concept of system, the behavior of any one manager can be seen as determined not only by his own personality needs and motives, but also by the way his personality interacts with those of his colleagues. Further, this relationship among organization members is also influenced by the nature of the task being performed, by the formal relationships, rewards, and controls, and by the existing ideas within the organization about how a well-accepted member should behave. It is important to emphasize that all these determinants of behavior are themselves interrelated.

For example, a typical manufacturing executive behaves in a certain manner not only because of his own personality, but also because his job as a plant manager requires him to have contact with a certain group of subordinates and with a number of executives at his own level as well as with a particular superior. While the individual personalities of these other executives may influence his behavior, our man will probably also behave as he does because of some expectations he shares with all these other managers about how a plant manager in this company should behave. The behavior of this executive will also be influenced by the fact that there is an established control system that measures certain costs and certain quality characteristics. The exact nature of the control system may be closely related to the nature of the technology. In a job-order shop a plant manager might be concerned about somewhat different matters from those that would confront the manager of a chemical processing plant. Both the formal organization and the technology may also be related to the shared expectations of how managers should behave, and because of all these characteristics the organization may attract managers with certain personality needs.

This description of an organization as a system has, for illustration, focused on the influences affecting the behavior of a typical manager. Our interest in this book, however, is in understanding the behavior of large numbers of managers in sizable organizations. This necessitates a central concern with two other important aspects of the functioning of systems. First, as systems become large, they differentiate into parts, and the functioning of these separate parts has to be integrated if the entire system is to be viable. As an analogy, the human body is differentiated into a number of vital organs, which are integrated through the nervous system and the brain. Second, an important function of any system is adaptation to what goes on in the world outside. We, as human systems, are very much concerned about dealing with the people and things that make up our external environment.

Differentiation and Integration and the Organization's Environment

It is on the states of differentiation and integration in organizational systems that this study places major emphasis. As organizations deal with their external environments, they become segmented into units, each of which has as its major task the problem of dealing with a part of the conditions outside the firm. This is a result of the fact that any one group of managers has a limited span of surveillance. Each one has the capacity to deal with only a portion of the total environment. If we take as an example either a division of a large, diversified corporation or a medium-sized manufacturing firm, we readily observe sales, production, and design units, each of which is coping with a portion of the organization's external environment. The sales unit faces problems associated with the market, the customers, the competitors, and so on. The production unit deals with production equipment sources, raw materials sources, labor markets, and the like. Such external conditions as the state of scientific knowledge and opportunities for expanding knowledge and applying it are in the most general sense the purview of the design unit. These parts of the system also have to be linked together toward the accomplishment of the organization's overall purpose. This division of labor among departments and the need for unified effort lead to a state of differentiation and integration within any organization.

The concepts of differentiation and integration as applied to organizations are not novel. They have been discussed in one way or another by a number of organization theorists and researchers. The best way to understand the differences in our approach from those of the other writers on organizational topics is to point to the way these concepts have been used in the past. In so doing we acknowledge our intellectual debt to these earlier theorists, and also recognize that the limitations of their use of these concepts have become apparent only through increased knowledge about the functioning of organizations.

Early writers about organizations, who have been labeled the classicists, were concerned with differentiation and integration.[3] Fayol, Gulick, Mooney, and Urwick dealt with how best to divide the tasks of the organization and how to obtain integration within it. From our point of view, however, their major failing was that they did not recognize the systemic properties of organizations. As a consequence, they failed to see that the act of segmenting the organization into departments would influence the behavior of organizational members in several ways. The members of each unit would become specialists in dealing with their particular tasks. Both because of their prior education and experience and because of the nature of their task, they would develop specialized working styles and mental processes.[4] For example, research managers and their subordinates tend to develop a distinct pace of work and orientation to time and to technical achievement as they spend hours puzzling over ambiguous problems. Similar points could be made about the ways of thinking and behavior patterns of other functional groups, such as production and sales. By differentiation we mean these differences in attitude and behavior, not just the simple fact of segmentation and specialized knowledge.

The observation of these differences is, of course, not new. In the course of this study, however, we have gone beyond everyday observation to identify three specific dimensions of the differences in ways of thinking and working that develop among managers in these several functional units. First, we have investigated the differences among managers in different functional jobs *in their orientation toward particular goals.* To what extent are managers in sales units concerned with different objectives (*i.e.*, sales volume) from those of their counterparts in production (*i.e.*, low manufacturing costs)? Second, we have been interested in differences in the *time orientation* of managers in different parts of the organization. Might production executives not be more pressed by immediate problems than design engineers, who deal with longer-range questions? Third, we have been concerned with differences in the way managers in various functional departments typically deal with their colleagues, that is with their *interpersonal orientation.*[5] Are managers in one part of the organization more likely to be preoccupied with getting the job done when they deal with others, while those in another unit pay more attention to maintaining relationships with their peers? In selecting these three categories of orientations to examine, we are not suggesting that these are the only differences that might be found among managers in different parts of an organization, but only that they seemed to be three important dimensions which our own observations of managerial behavior and earlier behavioral science research have suggested might be important.

The early organization theorists also did not recognize that each of the functional units would develop different formal reporting relationships, different criteria for rewards, and different control procedures, depending on the task of each unit.[6] Thus, the production department might have many levels in the management hierarchy, rewards for performance in meeting cost and quality standards, and control systems that measure these criteria in detail. On the other hand, a research unit might have fewer supervisory levels for each supervisor, rewards for performance of a broad objective such as "contribution to knowledge," and a much less precise control system. The variation in the *formality of structure* is the fourth dimension of differentiation among functional units that we have attempted to investigate.

In summary, when we refer to differentiation among units, we will mean differences in orientation and in the formality of structure. While these four characteristics are not all-inclusive and homogeneous, they do provide us with shorthand measures of differentiation as we define it in this study —*the difference in cognitive and emotional orientation among managers in different functional departments.* When we describe pairs of units or organizations as having more or less differentiation, we will be referring to whether the managers in the various units are quite different (more differentiation) in these four attributes or whether they are relatively similar (less differentiation).

As the early organization theorists did not recognize the consequences of the division of labor on the attitudes and behaviors of organization members, they failed to see that these different orientations and organizational practices would be crucially related to problems of achieving integration. Because the members of each department develop different interests and differing points of view, they often find it difficult to reach agreement on integrated programs of action.[7] A plant manager and a sales manager with different assigned responsibilities could quite naturally be expected to hold different views about the best price for a particular product. The plant manager might find a higher price, which would give him a wider latitude in production costs, desirable, while the sales manager might prefer a lower price, which would enable him to meet competition more effectively. This very elementary example of a built-in conflict of interests is compounded a hundred times over in a real organization, and the issues at stake are seldom so clear cut. It does, however, illustrate how we define integration—*the quality of the state of collaboration that exists among departments that are required to achieve unity of effort by the demands of the environment.* While we will be using the term "integration" primarily to refer to this state of interdepartmental relations, we will also, for convenience, use it to describe both the process by which this state is achieved and the organizational devices used to achieve it.

Conflict Resolution to Achieve Integration

While the early theorists did not explicitly recognize the relationship between the states of differentiation and integration, they did emphasize the need for integration in the organization. Their view, however, was that integration is accomplished through an entirely rational and mechanical process. If the total task of the organization was divided up according to certain principles, the integration would be taken care of simply by issuing orders through the management hierarchy, "the chain of command."[8] Our view, on the other hand, is that integration is *not* achieved by such an automatic process. In fact, the different points of view held by various functional specialists are frequently going to lead to conflicts about what direction to take. To achieve effective integration these conflicts must be resolved. The managerial hierarchy provides one means through which this resolution can come about, but it is not the only means. In many organizations integrating committees and teams are established or individual integrators are designated to facilitate collaboration among functional departments at all management levels. Routine control and scheduling procedures also provide a means of achieving integration. Finally, much integrating activity is carried out by individual managers outside official channels.[9] In our investigation we have also attempted to learn what factors have created the need for these various integrating devices, and what factors may be related to their serving a useful function in resolving conflict and achieving integration under various external environmental conditions.

In learning what factors determine the effectiveness of these various integrative devices, we have focused on another shortcoming in the early organization theories. Because of their notion that the process of achieving integration is mechanical and entirely rational, the early writers ignored the feelings and emotions connected with the achievement of organizational collaboration. As a consequence, they did not concern themselves to any great extent with the interpersonal skills required to achieve integration.[10]

This issue of interpersonal skills and their relationship to organizational integration has been a central concern of many later theorists and researchers, particularly behavioral scientists. These students of human relations have pointed to a variety of conditions in interpersonal relationships that are necessary to attain effective collaboration.[11] One such condition is that the parties who are dealing with one another learn to be open and frank about their positions as they work together. This openness leads to a climate of trust among the parties, which results in more effective problem solving. Closely related to this is the idea that conflicts must be confronted and brought into the open, rather than suppressed through the power of one side or avoided by the tacit consent of all.

While we have attempted to build on these and related ideas, they do have two major limitations. First, these theorists and researchers have been so concerned with problems of achieving collaboration that many of them have overlooked the equally important need for differentiation within organizations.[12] As a consequence, they appear at times to view recurrent conflict and disagreement as an avoidable dissipation of human energy. It is our view, given the need for differentiated ways of working and points of view in various units of large organizations, that recurring conflict is inevitable. The important question which we have tried to answer is how the specifics of each conflict episode can be managed and resolved without expecting conflict to disappear. In other words, how can integration be facilitated without sacrificing the needed differentiation?

Environmental and Task Attributes

These are some of the essential ways in which our approach differs from the ideas of the classical theorists and from those of many of the human relations researchers. There is, however, one other difference—to us the most important of all. Until very recently, as we have suggested, organization researchers and theorists have tended to view the internal functioning of effective organizations as if there was one best way to organize. No attention was devoted to the problem in which we are interested—that different external conditions might require different organizational characteristics and behavior patterns within the effective organization.

Our conviction that this approach might be fruitful came from several sources. First, our observations in many organizations as we gathered case studies for teaching provided concrete examples of the fact that different types of organizations were effective under different conditions. A case study of a highly profitable fine-grade paper company revealed an organization with very little reliance on formal rules and procedures, and with only limited control at the top of the organization.[13] Many important decisions were reached at the lowest levels of management. On the other hand, a case study of a highly profitable meat packing firm uncovered a different type of organization.[14] Here we found highly detailed procedures, many levels in the hierarchy, and very strong and dominant control at the top. We were intrigued with the question of what differences in the production technologies and in the markets might be related to the different styles of these two highly effective organizations.

Our curiosity was further stimulated by the involvement of one of the authors in a recent research study of worker response to varying technologies.[15] Although this particular research was not designed to examine overall organizational patterns in relation to technological and market differences, as the data were gathered and the researchers were exposed to eleven different organizations in as many industries it became apparent that there were differences in organizational patterns among these firms. If there was one best way to structure and administer an organization, how could these companies, which by economic criteria were all reasonably effective, have such diverse organizational styles?

Two recent studies have more systematically addressed these questions and further stimulated our interest.[16] Joan Woodward reported that successful organizations in different industries with different technologies are characterized by different organizational structures. For example, she found that the most successful firms in industries characterized by a unit or job-shop production technology had wider spans of supervisory control and fewer levels in the hierarchy than did successful firms in industries with continuous-process technology. A related finding was reported by Burns and Stalker in a study of firms in both a dynamic, changing industry and a more stable, established industry. Organizations in the stable industry tended to be what the authors called "mechanistic." There was more reliance on formal rules and procedures. Decisions were reached at higher levels of the organization. The spans of supervisory control were narrower. On the other hand, effective organizations in the more dynamic industry were typically more "organic." Spans of supervisory control were wider; less attention was paid to formal procedures; more decisions were reached at the middle levels of the organization, etc.

Both of these studies suggested that differing technical and economic conditions outside the firm necessitated different organizational patterns within it. More specifically these studies, particularly that of Burns and Stalker, suggested that the certainty of information or knowledge about events in the environment was one external dimension that impacted on the organizational variables in which we were interested. As we have already suggested, it might affect the organizational practices within departments, but it might also create different requirements for the way integration was achieved in the total organization. In more certain environments conflicts might be resolved and integration achieved at the upper levels through the management hierarchy. In less certain environments conflict resolution and the achievement of integration might have to take place at the lower levels of the hierarchy.

The questions raised by these two studies and our own observations of organizational situations can be summarized as follows:

1. How are the environmental demands facing various organizations different, and how do environmental demands relate to the internal functioning of effective organizations?
2. Is it true that organizations in certain or stable environments make more exclusive use of the formal hierarchy to achieve integration, and, if so, why? Because less integration is required, or because in a certain environment these decisions can be made more effectively at higher organizational levels or by fewer people?
3. Is the same degree of differentiation in orientation and in departmental structure found in organizations in different industrial environments?
4. If greater differentiation among functional departments is required in different industries, does this influence the problems of integrating the organization's parts? Does it influence the organizational means of achieving integration?

As we considered these issues, we realized that many organizational theorists and researchers were asking the wrong question. There probably is no one best way to organize. If our observations were correct, better practicing managers, consciously or by natural selection, recognized this fact as they designed and administered organizations under different environmental conditions. If we could investigate and compare organizations in several different environments, we might provide a systematic understanding of what states of differentiation and integration are related to effective performance under different environmental conditions, and further, we might learn something about how these states are achieved.

The Effective Organization and the Individual

While the matter of systemic differentiation and integration is our central interest, we are not ignoring the consequences of different organizational and administrative patterns for the satisfaction and development of individual managers. In fact, there seems to be an important connection for the individual between working in an organization structured to deal effectively with its task and his feelings of personal satisfaction and growth. Organizations so structured that members can deal realistically and effectively with their tasks will provide powerful sources of social and psychological satisfaction.[17] This is so because individual managers bring to an organization several motives that they seek to fulfill. Among the most important are a need for achievement, a need for affiliation, and a need for power.[18] To some degree individuals desire to accomplish something (need for achievement) by engaging in a task that they can perform effectively. For many managers in our modern society this would appear to be a dominant need that, for the most part, can be satisfied only within the organizational setting. Individuals also have a need for contact with others (need for affiliation). While many managers meet this need to an important extent in family and social relationships outside work, it must also be satisfied to some extent on the job.

Individual managers also have a desire to exercise some control over others (need for power). Again the need is met to some degree outside the organization, but it is also a motivating force in the organization. We should emphasize that while individuals vary in the intensity with which they feel these various needs, to some extent all organization members will seek gratification of their need for achievement in their work.

To satisfy this need, the organization must be structured to provide its members with an opportunity to do their individual jobs well. The importance of this is emphasized by the work of Robert White in relating the idea of competence to individual development and mental health.[19] He points out that individuals must feel a sense of competence in dealing with both objects and others in their personal environment. To borrow his phrase, "competence becomes in the course of development a highly important nucleus of motivation." [20] This underlying desire for mastery of one's personal environment seems closely related to McClelland's concept of need for achievement. As an individual satisfies the need for achievement he demonstrates competence in the world in which he lives. Making this connection helps us to understand that, if the individual is to satisfy this need, he must work in an organization that is suited to the requirements of the task he is interested in performing. Thus, if we concern ourselves with understanding what type of organization meets different environmental demands, we will also be confronting the question of developing organizations that offer a high probability of satisfying these basic needs of individuals for achievement and competence. While most members seem to have a high need for achievement, others may have stronger power or affiliation needs. Organizations well designed for the demands of their environment also can provide ample opportunity for the satisfaction of these other needs.

Research Strategy

A Comparative Approach

To find answers to our major question, we made a comparative study of competing organizations in each of several industries. In this way we have attempted to learn how the effective organization in a particular industry differed from its less successful competitor, and also to learn how effective organizations performing in different industries differed from one another. This work was carried out in two distinct, but related, phases. The first was a detailed study of a number of firms operating in one industry. This gave us an opportunity to sharpen our questions and find some answers about how the internal organizational states of differentiation and integration were related to each other and to effective performance in meeting the demands of a single industrial environment. The second phase was a study of a highly effective organization (by conventional economic and commercial standards) and a less effective competitor in each of two other industries. From this phase we gained more knowledge about the differences between effective and less effective organizations in other environments and, more importantly, we learned how differences in internal organizations were related to differences in external environments.

We shall not go into the details of what industrial environments were selected for study and why, except to state that in selecting the three industries for study we attempted to find ones that would differ in two important characteristics. First, and most important, we tried to examine organizations in industries with different rates of technological change in both products and processes. To do this is important, because industrial organizations that are going to survive in the future will undoubtedly have to deal with more and more rapid technological innovation. To learn

how the organization of the future might have to be shaped, we selected for study one industry currently dealing with rapid technological change and compared the organizations in this industry not only with one another, but also with firms currently operating in more stable, less dynamic industries. We also chose industries in which the dominant demands seem to come from different sectors of the environment. In this way we have attempted to learn something about how organizations that receive dominant pressures from the market differ from those confronted with dominant forces emanating from technical or scientific problems. Using these criteria we selected for study six organizations in the business of developing, marketing, and producing plastics materials; two in the container industry; and two in the consumer food industry.

Research Methods

To gather our information on internal organizational functioning we used both questionnaires and interviews, which were given to between 30 and 50 upper-level and middle-level managers in each organization. For example, in organizations that were the major product group of a larger diversified organization, the executives interviewed and questioned ranged from the product group or division general manager and his chief functional subordinates (*e.g.*, research director, production manager, sales manager) down to department managers and engineers in plants, scientists and engineers in research and development facilities, and field salesmen.

Data on the three industrial environments in which the companies operated were also gathered by means of both interviews and questionnaires. For this purpose, however, information was solicited from only the top executives in each organization (*e.g.*, the president, division or product group manager, and his principal functional subordinates), who would be in a position to be extremely knowledgeable about conditions in their particular industry. Additional details of the research methods used for gathering data about these organizations and their environments are described in the Methodological Appendix for those readers who are interested.

Outline of the Book

Chapters II through VI will describe the findings of this study and the theoretical framework they suggest. For several reasons, we will begin with the organizations in the plastics industry. First, this industry is characterized by rapidly changing technological and market conditions, and the organizational issues for firms in it may be typical of those to be faced in the future by firms in other industries. Related to this, the organizations in this environment are required to achieve both greater differentiation and greater integration. This means the organizational issues to be dealt with are more difficult here than in the other industries studied. Finally, because of these facts, we have examined six organizations in this environment. Therefore our comparison of the relationship between organizational and environmental variables in these organizations is more complete and more detailed than that in the container or food companies. For all these reasons it seems important to describe our findings in the plastics industry first, as a backdrop against which to compare the organizations in the container and food industries.

After the findings and the theoretical framework have been outlined, Chapters VII through IX will be devoted to a discussion of their implications for the practice of administration and the increased understanding and clarification of organization theory. In this section we will attempt to make use of the findings of this study to clarify the ideas of both the classical and human relations students of organization. We will also relate our findings to a selection of other contemporary studies of organizations that are developing parallel ideas. We will then apply the increased understanding of organization theory provided by these ideas to some of the pragmatic issues confronting managers in designing organizations and in developing the effectiveness of their members. Finally, we will attempt to examine the broader significance of these ideas for dealing with the problems of technological change and other emerging issues in society which will be of increasing concern for managers of the future.

CHAPTER II

Organizations in a Diverse and Dynamic Environment

IN THIS CHAPTER we will examine the internal functioning of six organizations operating in the same industry—a diverse and dynamic environment.[1] Later, we shall compare our intensive look at these organizations with an examination of two organizations in each of two other industrial environments with slower rates of technological and market change. Several features seem to distinguish what are generally considered to be the dynamic and growth industries from other business environments. These industries are usually based on a continually emerging body of scientific knowledge. It follows also that there is generally a great deal of uncertainty about the impact on markets of the new scientific and technological knowledge. Finally, these dynamic market and scientific conditions mean that the dominant competitive issue for firms in these industries is the capacity to innovate in both processes and products.

In 1963, when this study was initiated, a number of industries had these characteristics and would have been suitable as a setting for this part of our study. The electronics and aerospace industries both provided many of the characteristics we were seeking. However, because marketing in these industries involved the special issues inherent in the government being the dominant customer, we decided against them. The plastics and pharmaceuticals industries also met these characteristics, and after some deliberation we chose to investigate the former. This was not because we saw the plastics industry as more characteristic of future industrial conditions than the others, but because it also met the practical requirement of having organizations that were willing to participate in the research.

The Plastics Industry Environment

The plastics industry could be defined as extending from the basic bulk chemicals from which the plastics compounds are developed to the final product in the multitudinous forms in which it reaches the hands of the consumer. For the sake of better comparability among organizations, we limited our study to firms that produced and sold plastics materials in the form of powder, pellets, and sheets. Their products went to industrial customers of all sizes, from the large automobile, appliance, furniture, paint, textile, and paper companies, to the smaller firms making toys, containers, and household items. The organizations studied emphasized specialty plastics tailored to specific uses rather than standardized commodity plastics. They all built their product-development work on the science of polymer chemistry. Production was continuous, with relatively few workers needed to monitor the automatic and semiautomatic processing equipment.

Before we could begin our examination, we had to learn what demands this environment placed on organizations within it. To answer this question we interviewed the senior executives in each of the six organizations studied and also asked them to complete a questionnaire that measured certain attributes of their industrial environment as they saw them.*

These executives indicated in interviews that the dominant competitive issue confronting them was the development of

* See Methodological Appendix for details of the interviews and questionnaire.

new and revised products and processes. The life cycle of products was dwindling. Without new products any single firm was doomed to decline. And since all firms were steadily introducing new and revised products and processes, the future course of events was highly uncertain and difficult to predict.

According to these executives, the most hazardous aspect of the industrial environment revolved around the relevant scientific knowledge. It was difficult at any point in time to say just what all the relevant knowledge on a given topic was and just how certain it was. Cause and effect relationships were not well understood. Further contributing to this lack of certainty was the long time required to get feedback about whether a particular scientific investigation had been fruitful. One executive put it this way:

> We very often know the performance characteristics required by a customer or for a new application, but as far as research is concerned, you might as well be asking for the perfect plastic. We have to be concerned with technical feasibility right from the start, and then you don't know as you are involved in a lengthy testing and development process.

The general manager of another organization further emphasized the uncertainties of scientific knowledge:

> The development of plastics materials is more of an art than a science. We often don't fully understand what is needed to meet customer requirements, and if we do, we don't know how we can process it. However, we have developed the art to a high enough degree so we can hit the target area, even if we can't hit the target in every case.

The research director in another organization stressed the lack of understanding of cause and effect relationships:

> One problem in the development of new applications is the identification of a need. Once we have identified this, we can easily establish the required product specifications. Of

course, achieving them is another thing. The problem here is that our knowledge is not nailed down as precisely as it is, say, in chemicals. Sure we are able to do a lot of things by trial and error, but this is solving problems without knowing the reason why.

As these comments suggest, all the top managers interviewed felt knowledge about the market was more certain than scientific knowledge. Customer needs and other required information could be specified at any particular time. Causal relations were better understood. For example, the result of a price change could be predicted with some accuracy. Furthermore, definite feedback about the results of some action in the market was available quickly, often within a month. Yet there were a number of uncertainties, many stemming from the great variety of customer needs. There were also problems in specifying all the variables that must be taken into account. As one executive said:

> The market we sell to is broad and diverse. For example, in the toy business if the product comes out in the right color at the right price, anybody will buy it, and anybody can make it. In contrast, for wire and cable customers, the right technology, service, and delivery are critical, and price is at the end of the line. Usually some minor technical property sells your product in that business. It is very rare that you get exact duplication from anybody.

Another manager made the same point:

> Because our customers typically use the products we sell in a chemical reaction, we have a relatively high level of control over the suitability of our product to the customer. In other words, how the customer uses the product affects its utility to him. Consequently, we have a hundred markets, each different in requirements because of the customers' processing needs.

While scientific knowledge was highly uncertain, and the market moderately uncertain, technical and economic factors, according to these executives, were more predictable. Once the product specifications had been developed, the technical and economic knowledge needed to carry out the manufacturing process were highly certain. At this stage the desired product characteristics were well defined from both a technical and a cost standpoint, and production personnel had particular targets to shoot at. Highly reliable information could be obtained about how well the process was operating. The devices necessary to monitor the major production variables—*e.g.*, pressure, temperature, and chemical composition—were often built right into the production process. Causal relationships among these variables were also relatively well defined. Finally, the time required to get clear information about the consequences of actions in this area was short—at most only a matter of hours or days. As one executive commented:

> In production, life is really plain, as it is geared to running the kind of plant and equipment which they currently have and where most of the decisions are built in.

While what we call "environment" is fairly self-evident as applied to research and marketing, the use of this term in relation to production requires some explanation. Contrary to conventional usage, we have chosen to conceive of the physical machinery, the nonhuman aspect of production, as part of its environment. Production executives must draw information from this equipment's performance and analyze it in terms of costs, yields, and quality, just as they must also draw information from outside the physical boundaries of the firm about newly available equipment and alternative processes. It is this information and knowledge from all these sources that we are interested in characterizing as certain or uncertain. Readers who find this awkward may prefer to think in terms of "the production task" rather than of the "techno-economic environment."

In short, what the top executives in these organizations told us about the environment of the plastics industry was that different phases of it vary in degree of certainty. One manager summarizes their conclusions very well:

> The requirements in this industry are that you must have good and continuing technology, sound economics in your production, and you must market your product at the lowest costs. I'd say these are all equal. But first among equals, and the most critical, are the research and marketing areas. I would say that they are the most critical and the most difficult because there are so many variables involved. You know what you are doing in production. You have the facts there.

The data collected in our questionnaires generally confirm the conclusions we reached through the interviews (Table II–1. The clarity of information, the uncertainty of cause and effect relationships, and the time span of definitive feedback scores have been combined to get a total uncertainty score.* While the differences in the uncertainty scores for different parts of the environment are not highly significant,

* The justification for combining these scores is based on both their intercorrelation and their conceptual relationship. While there were some significant intercorrelations (clarity of information with certainty of causal relationships in the scientific sub-environment and with time span of definitive feedback in the market and scientific parts of the environment at .05; timespan of feedback with certainty of causal relationship in the scientific part of the environment at .01), these were not perfect. However, as Victor H. Vroom, *Some Personality Determinants of the Effects of Participation* (Englewood Cliffs: Prentice-Hall, 1960), p. 25, has pointed out, the issue of how homogeneous items should be before they are combined is a complex one. High intercorrelation reflects only high reliability and considerable overlap between questions. Low homogenity may be due to either unreliability among items or to the fact that the items measure different things. He concludes by pointing out, "To the extent that these items are conceptually related and represent variables which have similar effects, combination of items into a single score will broaden the range or breadth of the resultant measure." In our case the three components of certainty were conceptually related and were measuring similar but not identical effects. We have therefore felt justified in combining them into one certainty score to obtain a broader measure.

the order is clearly in the expected direction, indicating, as do the interview data, that scientific knowledge is least certain; market knowledge, next; and techno-economic knowledge, most certain.

Table II–1

RELATIVE UNCERTAINTY OF ENVIRONMENTAL SECTORS [a]

Environmental sector	*Clarity of information*	*Uncertainty of causal relationships*	*Time span of definitive feedback*	*Total uncertainty score*
Scientific	3.7	5.3	4.9	13.9
Market	2.4	3.8	2.8	9.0
Techno-economic	2.2	3.5	2.7	8.4

[a] Higher scores indicate greater uncertainty.

One slight disparity between the data gathered in questionnaires and the impressions we gleaned from our interviews concern the relative uncertainty of the market and techno-economic information. In interviews, the managers indicated that the market was considerably the less certain of the two. While their responses to questionnaires tended in this same direction, the differences were not so great as the interviews led us to expect. The reason seems to lie in the phrasing of the questionnaire, which focused more on the task of the sales unit itself and less on the nature of the market. Since much of the market uncertainty was handled by groups other than the line sales department (*e.g.*, development departments, product integrators), the task of the sales group was seen as more certain than the oral descriptions of the market suggested. This seemed to be particularly true with respect to the clarity of information about the market. On this point, therefore, we have relied more on the evidence gathered in interviews than on responses to the questionnaire.

From all these data we concluded that the differences in

uncertainty among these three aspects of the environment were such that the primary task of each major functional department was essentially different. As we suggested in Chapter I, this would mean that different orientations and organizational practices would tend to develop within each of the basic departments. We will review below our specific hypotheses about these departmental differences and compare them with our empirical findings.

Departmental Differentiation

Basic Departments in Six Organizations

Each of the six plastics organizations studied was a major product segment of a large, diversified chemical company. While there were some differences in their products, they all competed in the manufacture of basic plastics materials for sale to industrial customers. Each organization was a self-contained entity, carrying out a complete cycle of designing, marketing, and manufacturing. In general, the only constraints imposed by the corporate headquarters were at the broadest policy level, where decisions might affect more than one product segment of the corporation.

Each of these organizations had four *basic functional departments*—sales, production, applied research, and fundamental research. The functions of the sales and production departments were already evident from what we have said about the nature of the environment. The two research units dealt with different phases of the scientific segment of the environment. As the title implies, fundamental research departments were generally supposed to relate to the less certain aspects of relevant scientific knowledge. Although all the applied research laboratories dealt with the more certain scientific problems, their assigned missions varied as we shall see later.

In examining these four basic departments, we will first summarize how earlier behavioral science researchers had suggested that they should be differentiated in order to deal effectively with the differing degrees of certainty in their respective parts of the environment.[2] This will take the form of a description of particular required internal characteristics of each department. Second, we will report our empirical findings on the *actual* internal characteristics. This will tell us to what extent these departments were consistently different from one another in each predicted dimension.

Formality of Structure, Predicted and Actual

The first dimension we examined was *formality of structure*. Did the basic departments show differences in formal structure that could be related to differences in the certainty of their part of the environment? We expected, from earlier behavioral studies, that the research units, dealing with highly uncertain tasks, would be less formally structured than the production units. Organizational practices and procedures would, we thought, be less controlling and constraining than in production departments, because the task would require a setting where scientists and engineers could openly and freely explore scientific problems with one another. On the other hand, we expected that production managers, to do their jobs effectively, would require more established routines and tighter controls. The task of the sales departments, which fell between the other two in certainty, would, we assumed, require a departmental structure between the extremes of the other units.

We collected our data on formality of structure by examining organizational manuals and charts and by interviewing each department manager about the practices in his unit.* These data indicate that the differences among departments

* For a more complete description of this and later data collecting methods, see the Methodological Appendix.

were in a direction that could reasonably be related to environmental certainty. The production units have the most formalized structure in four of the six organizations. In the other two, the production unit has either the next to highest formality of structure or was tied for highest. Compared with the other departments in each organization, the production units had more levels in the managerial hierarchy and a higher ratio of supervisors to subordinates, as well as more frequent and more specific reviews of both the department's and individual managers' performances. Precise data for judging performance were quickly available and readily used. Similarly, since the manufacturing units' activities could largely be programmed in advance, these units relied more on formal rules and procedures than did the sales or research units.

At the other end of the spectrum were the fundamental research units. In four of the organizations this laboratory had the least formal structure, and in the other two it was tied for the least. Specifically, more subordinates reported to each supervisor, and there were fewer levels in the managerial hierarchy. Performance reviews were general and infrequent, and there was little reliance on written rules or procedures.

The structures of the applied research laboratories and the sales departments fell between the extremes represented by production and fundamental research. In general, the sales units were more structured than the applied research laboratories. This, too, is consistent with relative environmental certainty. These data, then, strongly indicate that there were differences in the structural features of the various departments in each organization, and that these differences were related to the nature of that phase of the environment with which each unit was dealing.

Interpersonal Orientation, Predicted and Actual

The second dimension along which we measured differences among departments was the kinds of interpersonal relationships among department members. Prior research had suggested that the members of different departments would find it necessary, if they were to be effective, to develop different interpersonal orientations related to the nature of their task. Specifically earlier work had suggested that members of units engaged in either highly certain or highly uncertain tasks would develop task-oriented interpersonal styles. Members of units with moderately uncertain tasks would develop relationship-oriented styles.[3] Thus, production engineers, plant managers, and other manufacturing personnel whose tasks were very certain would generally, we predicted, prefer interactions aimed primarily at getting the job done and would not be particularly concerned about maintaining social relationships. Sales and marketing managers, who were dealing with a moderately uncertain task and who were also accustomed to being concerned with customer relations, would be expected to care more about fostering positive social relationships among their co-workers. Research scientists and engineers were expected to fall between the production and sales extremes, with the fundamental scientists preferring contacts which emphasized task accomplishment. The high uncertainty of their jobs, which were often carried out individually, would lead them, we predicted, to be less concerned with social relationships than applied research personnel who had more certain tasks.

To measure interpersonal orientations we used a standard paper-and-pencil instrument, which indicates a person's preference for relationship-oriented or task-oriented interactions. The data collected generally supported our predictions. In two of the six organizations the sales departments were the most relationship-oriented, and in the other four they were the second in this respect. Production personnel indicated a preference for task-oriented relationships. They were the most task-oriented in four of the organizations and were tied for second in a fifth. (Production personnel in the sixth organization deviated from this pattern, for reasons to be explained shortly.) Since the production procedures and controls and the processing technology provided the necessary coordination within the department, production personnel generally did not see positive social relationships as essential. If these managers found it difficult to work with a particular colleague, their attitude was to get on with the job without special regard for maintaining positive relationships. Both fundamental and applied research personnel fell between the extremes of production and sales in their interpersonal orientation. The applied people having a moderately uncertain task leaned toward the relationship orientation of sales and the fundamental researchers tended more toward the task orientation of production. Thus these differences in orientation were also generally consistent with our predictions drawn from earlier research.

Time Orientations, Predicted and Actual

We expected that members of these four basic departments, because of environmental differences, would also develop different orientations toward time. Sales and production personnel, dealing with problems that often provide rapid feedback about results, would, we assumed, have their attentions focused on short-term matters. Research scientists and engineers would, we thought, have longer-range concerns, because tangible results of their efforts could be judged only when they had solved the scientific and technical problems leading ultimately to a new product or a new process. People working in these research settings would need to be comfortable in delaying the gratification they received from feedback about the results of their work. As with any of these attributes, in describing the required time orientations we are obviously only describing our predictions about a general tendency in the indicated direction. Certain managers within the manufacturing unit may be required to be concerned about longer-term operations, and certain research personnel may be working at solving a particular customer problem where success would be evident within a few weeks.

The time orientation was measured by asking managers, in the questionnaire, how much of their time was devoted to activities that contributed to company profits in different future time periods—one month or less; one to three months; three months to one year; and one year or longer. Here again, we found support for our prediction about the relationship between the nature of each part of the environment and the orientation of departmental personnel. The primary time orientation of members of each department was related to the time span of definitive feedback. The members of both the production and sales departments reported that most of their work dealt with matters that would affect profits relatively soon, often less than one month in the future. This was consistent with the rapid feedback available from both the market and the techno-economic portions of the environment. In contrast, administrators and scientists in the fundamental laboratories were oriented primarily toward much longer time horizons, often several years in the future. Again this was consistent with environmental demands, since the final evidence of success or failure in their activities was available only when a project was ended, which usually required a year or longer. While the members of the applied research laboratories in most of the organizations also focused on longer-term objectives, there were some exceptions that gave the total group a medium-term to long-term orientation. These exceptions seemed to grow out of differences in the assigned activities of these units, which we will discuss shortly.

Figure II–1 summarizes the data on the three dimensions

we have examined so far—formality of structure and interpersonal and time orientations—for the basic departments in all six organizations. It suggests quite clearly that across all organizations there were differences among departments that were consistent with our predictions. It should be emphasized, as indicated above, that structure and time orientation ap-

Figure II–1

DIFFERENTIATION RANK OF DEPARTMENTS IN SIX ORGANIZATIONS [a]

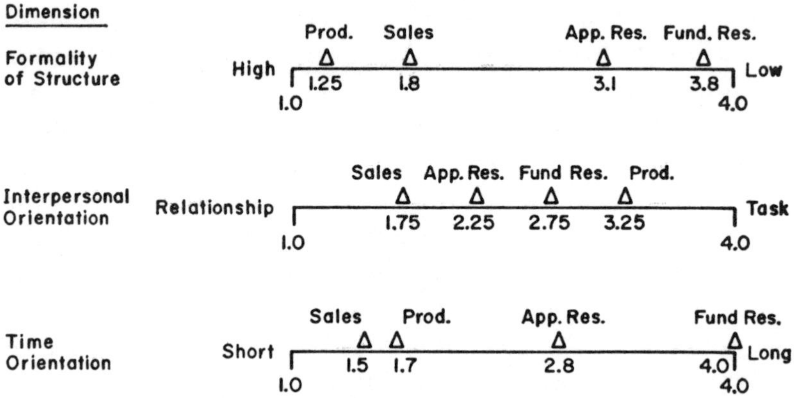

[a] Numbers are the average rank order for all six organizations of the basic departments along the specified dimension within each organization.

pear to vary in almost a linear relationship to the certainty of the task, while interpersonal orientation has more of a curvilinear relationship to certainty. When the task is either very certain or very uncertain members prefer task-oriented interactions, but when the task is moderately certain they rely more on relationship-oriented interactions.

Goal Orientation, Predicted and Actual

The fourth, and probably most obvious, way in which we expected these departments to differ is in the orientation of their members toward goals and objectives. This attribute is not related to the certainty of the parts of the environment, but simply to the fact that each of the functional units had been assigned the job of dealing with a particular aspect of the total environment. If these managers were to be effective in doing their specialized tasks, they must focus their attention clearly on objectives and goals directly related to them. Sales managers must be concerned with accomplishing market objectives. Manufacturing managers must pay attention to such techno-economic goals as processing costs, raw materials costs, and the quality of the finished product. Research personnel must be primarily concerned with both the development of new scientific knowledge and its successful application to products and processes. In each case we were interested to see whether the members of each unit had their attention tightly focused on the specific goals of their department, or whether their concerns were more diffused. Data to measure this orientation were collected on the questionnaire by asking managers to indicate their concern with nine different decision-making criteria—three each dealing with the market, techno-economic factors, and science—*e.g.*, the probable response of competitors to the decision (the market); the effect of the decision on contributions to scientific knowledge (science); and the plant facilities required by the decision (techno-economic).

Our prediction in this area also turned out to be largely true. Sales personnel in all six organizations indicated a primary concern with customer problems, competitive activities, and other events in the marketplace. Manufacturing personnel were all primarily interested in problems of cost reduction, process efficiency, and similar matters.

In the research laboratories, however, we did not find so strong an orientation as we had expected toward scientific objectives. In three of the organizations fundamental scientists did indicate a primary concern with the development of new knowledge, but in the other three they were more concerned with such techno-economic factors as technological improvements for cost reduction and quality control. A similar orientation was reported in all but one of the applied research units. Although we had anticipated that these personnel would be strongly attuned to scientific goals, in retrospect this result is not too surprising because much of the activity in these laboratories, particularly the applied ones, was devoted to process improvements and modifications. In any case, in those research units where scientists and engineers indicated a primary orientation toward techno-economic goals they also indicated a strong secondary concern with scientific factors.

At this point it is important to review briefly the arguments and evidence that we have so far presented. Based on our prior evidence about the environment of the plastics industry, we expected to find variations in the amount of uncertainty present in the different environmental sectors. Both the interview data from key executives and our measurement of uncertainty factors supported this prediction. We further predicted that the basic departments in each organization would tend to be different from one another along four important dimensions. We have presented data that clearly show that this was, in fact, true. The departments were differentiated in ways that were predicted, relative to environmental attributes. While the existence of these differences will probably come as no surprise to most businessmen, it not only confirms general impressions but, more importantly, provides an explanation of why there are differences among basic departments. We must emphasize, however, that these findings indicate only a general tendency toward the predicted pattern of differentiation, and that it is still problematic how precisely the pattern in any one organization will fit its environment. The key question is: Were the organizations that have achieved the closest fit between departmental differentiation and the attributes of their environment also the high performers in terms of economic results? Having prepared the way to address this question, we will start by examining data about the performance of these six organizations.

Differentiation and Performance in Six Organizations

Organizational Performance

We did not, we must stress, pre-select these organizations on the basis of their performance, since the required information was not public. We did, however, try to choose organizations with a general reputation for being either highly successful or somewhat less effective. As it turned out, based on the measures of performance we obtained, we did achieve a balance of highly effective organizations and those with problems in achieving long-term success. The six organizations were ranked in terms of three performance measures: change in profits over the past five years; change in sales volume over the past five years; and new products introduced in the past five years as a percent of current sales (Table II–2).* The total ranking of these variables gave us one measure of performance, but we felt it was also necessary to check this index

* While ideally we would have preferred to use as measures of performance such indices as return on investment, or profits as a percentage of sales, the managements considered these to be so confidential that they would not make the information available to us. We were able to obtain the growth rate of profits over the past five years (a measure of past profit performance); the growth rate of sales over the past five years (a measure of past performance in increasing the size of operations); and an index of new products introduced in the past five years as a percent of current year's sales (a measure of future potential performance). Thus, these indices were measuring different dimensions of the organizations' performances, and we were not predicting that they would necessarily be closely related (in fact, they were not significantly interrelated). These three dimensions have been added to get at least one crude empirical measure of some of the salient dimensions of the relative effectiveness of these organizations in coping with their environment.

Table II–2

ORGANIZATIONAL PERFORMANCE

(Ranked from High to Low Performance)

Organization	*Change in profits over 5 years*	*Change in sales volume over 5 years*	*New products developed in last 5 years as % of current sales*	*Total ranking* [a]	*Managers' subjective appraisals* [a]	*Performance rating*
High A	2	3	1	6(2)	2½	High
High B	1	1	3	5(1)	2½	High
Medium A	3 [b]	2	4	9(3)	1	Medium
Medium B	6	4	2	12(4)	4	Medium
Low A	4	6	6	16(6)	6	Low
Low B	5	5	5 [c]	15(5)	5	Low

[a] Spearman's rank-order correlation between the ranking of index totals (in parentheses) and the ranking of managers' subjective appraisal was significant at .05 (corrected for ties).

[b] Organization medium A had been operating at or near the break-even point during this period. A small increase in profit made this index rise unrealistically in relation to the other organizations, so the average of the other two indices was used.

[c] Organization low B, which had been in existence only five years, could only report that all products had been introduced in the past five years, so the average for the rankings of the other two indices was used.

by obtaining from the chief executive in each organization a subjective appraisal of how well his organization was doing as compared to what he would like. The rank order of this subjective appraisal correlated closely with our total performance index, which gave us some assurance that we were obtaining at least a crude measure of relative performance.

As a further check on these performance measures, we also interviewed the top two or three executives in each organization to learn their views about how well each company was doing. The information gathered in these interviews, which was consistent with that reported in Table II–2, also suggested that the six organizations did cover a wide range of performance.

In both high-performing organizations the top executives interviewed were highly satisfied with the way things were going. They were pleased with past and current performance, and felt that the future looked even more promising. Top managers in the medium-performing organizations were less enthusiastic in appraising performance. In medium A the executives reported that they had been through a difficult period, but that in the current year the organization had begun an upward trend, which needed even greater improvement.* (This recent shift to a more favorable performance probably accounted for the fact that its chief executive rated this organization's performance closer to ideal than did the chief executive in any other organization. As Table II–2 indicates, this was the only important deviation of the top managers' appraisal from our index. This man's enthusiasm was understandable, but according to our evidence, not completely accurate.) The situation for medium-performing organization B was similar to that at organization A. Five to ten years be-

* The letter designations have been assigned to enable us to distinguish between the two organizations at each performance level. They are not intended to suggest that there was any significant performance variation between the two organizations at the same level.

fore this study was made it had been an industry leader. Its position had slipped, but at the time of the study its top management felt that the unfavorable trend had been reversed. Again, there was a feeling that there was room for further improvement, but also the belief that the performance trend was in the right direction.

The low-performing organizations were both characterized by their top administrators as having serious difficulty in dealing with this environment. They had not been successful in introducing and marketing new products. In fact, their attempts to do so had met with repeated failures. This record, plus other measures of performance available to top management, left them with a feeling of disquiet and a sense of urgency to find ways of improving their performance.

From these interviews, the top managers' subjective appraisals, and the empirical evidence available, we felt justified in dividing these six organizations into the three broad categories of performance that we have indicated. With this general understanding of how these organizations were succeeding in dealing with the plastics industry environment, we can now examine their internal functioning to learn if their state of differentiation was related to effective performance.

Differentiation and Performance

We have seen that the structures and interpersonal, time, and goal orientations of the functional units in all six organizations were generally different and that this differentiation was usually consistent with the requirements of their different sub-environments. Some units in some of the organizations, however, had not fully developed the attributes that we predicted would be required to deal with their part of the environment. This demonstrates the point made earlier that the achievement of the required state of differentiation was problematical. Organizational practices and orientations that fit the task requirements do not necessarily emerge in all departments. Why this fit occurs in some situations and not in others is a complex question that cannot yet be answered. We were, however, interested in learning whether the attributes of functional departments tended more to fit the requirements of their part of the environments in the high-performing organizations than in the less effective ones.

This, in fact, turned out to be the case (Table II–3). The units in the high-performing organizations deviated from their environmental requirements significantly less than did departments in either the medium- or low-performing firms. This suggests that the achievement of a degree of differentiation consistent with the requirements of the environment is related to the organization's ability to cope effectively with its environment.

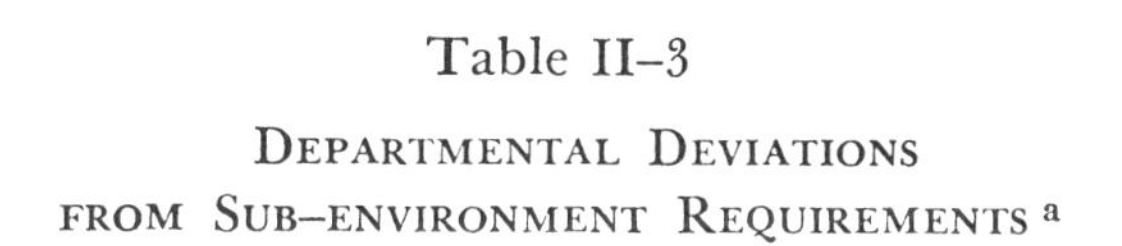

Table II–3

DEPARTMENTAL DEVIATIONS
FROM SUB–ENVIRONMENT REQUIREMENTS [a]

	Number of deviations
Two high-performing organizations	4 [b]
Two medium-performing organizations	10 [b]
Two low-performing organizations	10 [b]

[a] Deviations reported represent the number of deviations in all four attributes (structure, interpersonal, time, and goal orientations) for fundamental research, sales and production departments. The applied research departments were not considered because of the variations in their assigned tasks. (For further details, see the Methodological Appendix.)

[b] $P > .05$ (Fisher's exact test)

When personnel in a particular department share attitudes and interests that focus clearly on departmental goals and have time horizons consistent with their task, they will be more effective. Similarly, when they have developed interpersonal modes that are appropriate to the nature of their work, their performance should be effective. Organizational practices that provide a degree of latitude or control consistent with the nature of the task will also facilitate the department's dealings with its part of the environment.

Departmental Differentiation and Integration

So far we have focused on the fact that the functional departments in these six organizations had developed differentiated attributes that seemed to be related to their overall effectiveness. There is, however, another side to these differences in orientations and organizational practices. They also mean that when it becomes necessary to make joint decisions, managers from different departments will approach the problem from different frames of references and may have difficulty in collaborating effectively. We now want to learn whether there was a systematic relationship between the degree of differentiation between any two departments and their ability to achieve effective integration. We have already mentioned in general terms the requirements in this industry for close integration of the basic functional units, but we must now look more closely at these requirements to set the stage for testing the relationship between differentiation and integration.

Required Integration

The top executives interviewed about conditions in this environment indicated that the focus on innovation as a dominant issue created the requirement for two especially critical interdependencies among departments. One of them stated it this way:

> I would have a difficult time distinguishing where [between sales and research or between production and research] it is necessary to obtain the most coordination, but it does seem to me that there is a sequence which it is helpful to understand. First, it is necessary to have contact between research and the people in contact with the customers. They

> spend a long time together working on the initial stage of development. They work together deciding what the customer needs and in testing the product in the customer's shop. Finally, the production people are brought in, when the process is ready to be handed over to them. How soon they are brought in would depend upon whether equipment modifications would be necessary. The research people then have to work out the processing problems with them.

As this executive points out, the sales and research units in any organization in this industry were required to work closely together and to make complicated joint decisions. This tight connection was necessary because the problems of testing and introducing a newly developed or modified material required the combined efforts of marketing personnel and research scientists and engineers. The high degree of integration between the research and production units was necessary to inform research scientists about technical problems that they might be able to overcome; to acquaint production personnel with new processing techniques that could be introduced; and to determine what technological innovations were possible and desirable.

We have been discussing these relationships between production and research and sales and research units as if there were only one research department in each organization. Since each actually had two research units, this complicates the problem of establishing where close integration was required. This is especially so because of differences in the tasks assigned to the applied research unit in the six organizations. In some, the applied laboratories were conducting long-range research similar to that carried on in the fundamental laboratories. In others, the applied laboratories dealt with more certain problems, such as short-term process improvements and customer technical service. This meant that although the applied research units in all six organizations had to be closely integrated with both sales and production, there were some variations in which basic units were required to work closely with fundamental research. For example, in some organizations close integration was required between applied research and fundamental research, as well as between fundamental research and production, while in other organizations the only basic unit requiring close integration with fundamental research was production. In only one organization did the executives indicate a requirement that sales and fundamental research be closely integrated.

One other factor complicated determining the points within these organizations where close integration was required: All six organizations had established a group or department of "integrators," whose responsibility was primarily to facilitate collaboration among the middle and lower managers in the basic units. According to all of the top managers, close integration was required among these "integrating units" and all four basic units.

While these are the relationships where close integration of functional units is required, the top managers interviewed were not suggesting that the sales-production relationship was not important. Rather, the coordination required here could be more nearly routine, since it generally dealt with relatively nonproblematical operational issues of customer delivery and schedules.

The major interdepartmental relationships in which close integration was required in this environment are summarized in Table II–4. As we have already seen, these same units that had to work closely together also had to be quite different from one another. This poses two fundamental questions, which will interest us throughout the book. Do great differences among departments in ways of thinking and in organizational practice generate problems in achieving integration? If this is the case, how do effectively competing organizations obtain both the required state of differentiation and the required state of integration?

Table II–4

INTERDEPARTMENTAL RELATIONS REQUIRING TIGHT INTEGRATION IN THE PLASTICS ENVIRONMENT [a]

Integrating unit–	Sales
Integrating unit–	Production
Integrating unit–	Applied research
Integrating unit–	Fundamental research
Applied research–	Production
Applied research–	Sales

[a] In addition, different basic units required close integration with fundamental research in each organization. In analyzing data in regard to the fundamental research units, we have been guided for each organization by the definition of the requirement for tight integration given to us by its top managers.

Inverse Relation Between Differentiation and Integration

Based on earlier behavioral studies, as we suggested above, we predicted that there would be an inverse relation between the degree of differentiation between any two departments in a single organization and the quality of the integration between them. When we tested for this, we did in fact find in all six organizations a very clear inverse relationship between the magnitude of differences in orientations and structure between pairs of departments and the effectiveness of integration between them.* (This relationship is schematically represented in Figure II–2.) The more similar two departments were in structure and in the orientations of their personnel, the more effective was the integration between them. Conversely, the more different two departments were in these attributes, the more difficulties they were having in achieving

* Integration was measured by asking in the questionnaire for an evaluation of the state of interdepartmental relations. Response could vary on a seven-point scale, from "sound; full unity of effort is obtained" to "couldn't be worse; bad relations; serious problems exist." The other methods used to measure this relationship are explained in the Methodological Appendix. An example of the rank order of the degree of differentiation and the degree of integration for one of the six organizations is also presented.

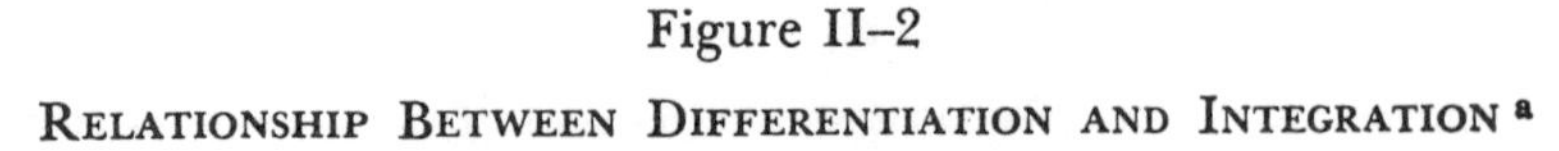

Figure II–2

RELATIONSHIP BETWEEN DIFFERENTIATION AND INTEGRATION [a]

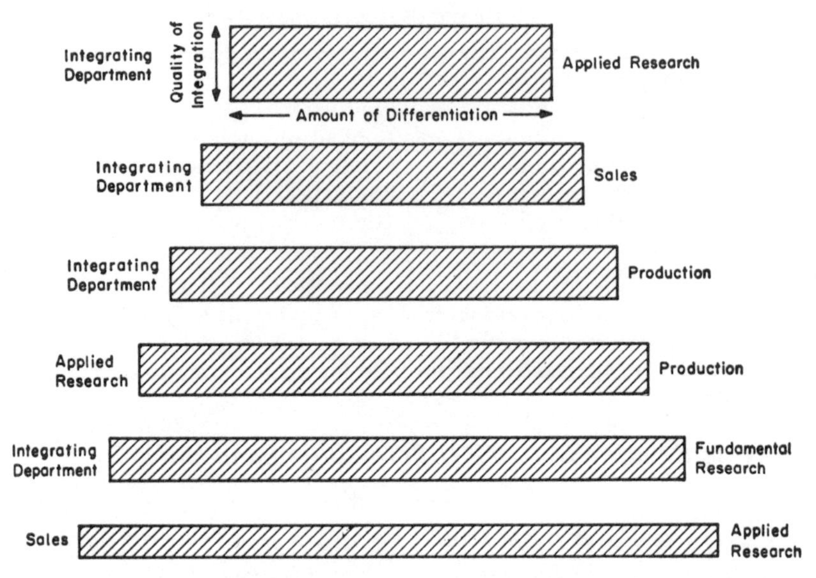

[a] This is a schematic representation of the relationship among departments in one organization. The longer the bar, the more differentiation; the wider the bar, the better the integration. This relationship held between pairs of units in all six organizations.

integration. (This relationship was statistically significant in five of the organizations at .01 and in the sixth at .05, using Spearman's coefficient of rank-order correlation.) In essence, the degree of differentiation among departments within each organization was antagonistic to the quality of integration obtained.

Given this finding, we could logically expect, in comparing the states of differentiation and integration among organizations, to find that the more highly differentiated all the units in the organizations were, the more difficulty the organization would have in achieving high-quality integration. If this were the case, we wondered if it would be at all possible for organizations simultaneously to achieve both the high degree of differentiation and the tight integration that the top managers of these organizations had told us were required for successful performance in this industry.

Differentiation, Integration, and Performance

To find an answer to this question we computed and compared the average amount of differentiation between departments in each of the six organizations and the average quality of integration achieved in each organization.*

The results were highly intriguing. The two organizations with the most successful performance records had, in fact, achieved the highest degree of integration of the six and were also among the most highly differentiated (Figure II–3). As we indicated above, the differentiation of the various units was more in line with the demands of the environment for these two organizations than for the others. Managers in both organizations also indicated in both questionnaires and interviews that the units in these organizations worked very well together, achieving the integration required by the environment. A typical view of these relationships is provided by the comments of a research manager in one of these two organizations, describing his unit's relations with the production department:

> We have had good relations with production lately. I have seen other situations where production people wouldn't follow research changes in processes. In [this organization] when we have the recipes outlined, the production people go over them with us and understand them and follow our procedures. At first we had some difficulties, probably because I had to learn what their capabilities were, but now they just sit down with us and we go through the problems and work them out and they follow our recipes.

* The procedure used to make this comparison is described in the Methodological Appendix.

Figure II–3

DIFFERENTIATION,[a] INTEGRATION, AND PERFORMANCE IN SIX PLASTICS ORGANIZATIONS

DIFFERENTIATION

INTEGRATION	High		Low
High	D = 9.4, I = 5.7, High A	D = 8.7, I = 5.6, High B	D = 7.5, I = 5.3, Medium A
Low	D = 9.0, I = 5.1, Medium B	D = 9.0, I = 4.9, Low A	D = 6.3, I = 4.7, Low B

[a] High differentiation score indicates high differentiation. High integration score indicates effective integration. Differences between integration scores are significant as follows:

High-performing and low-performing organizations significant at .01.

High-performing and medium-performing organizations significant at .05.

Medium-performing and low-performing organizations significant at only 10.

Meaningful comparisons of differentiation scores cannot be made.

Other comments from managers in both organizations described the state of integration between units in similar positive terms. All of this evidence indicates that the states of differentiation and integration in the two high-performing organizations were in line with the environmental requirements.

The two medium performers were not simultaneously achieving the required differentiation and integration. They seemed, in effect, to have traded off some differentiation for improved integration, or *vice versa*. Medium-performing organization A had relatively close integration of units, but its state of differentiation was very low. Its departments were able to work well together, but perhaps because they were not very much different. On the other hand, medium-performing organization B had achieved high differentiation, but apparently at the expense of various departments' being able to work effectively in reaching joint decisions. The integration scores from questionnaires and interview comments indicated that there were many problems in this area. For example, two research managers described their problems in obtaining interdepartmental collaboration:

> I think the biggest problem is communications with [the integrating unit] so we can know the needs. You need to know what sales and production needs are. They have customer complaints. If we knew the customer complaints, we could help. In this organization there isn't good communications linkage on these things.
>
> * * * * *
>
> Cooperation at times with production hasn't been too good. At times we are looked on as outsiders, even though they are running experimental processes we developed. They do it without consulting us. They may get general ideas on a run before they start, but once it is started, there is no suggestion or discussion. I don't know whether they don't bother or they don't want our help.

Low-performing organization B had units that were very similar in orientation and structure, which also were unable to attain effective integration. This clearly fits our hypothesis that a lack of differentiation and integration would contribute to low performance and presents the clearest example of this condition.

The other low-performing organization, A, while it had units that were well-differentiated, was achieving the second lowest degree of integration. For example, sales and research managers in this organization discussed their problems in this fashion:

> I don't feel that sales is brought into product evaluation early enough. My feeling is that if we were brought in earlier, we could save a lot of time that research spends working on things. The way things are set up now, we are absolutely independent of what is going on with new products. Our only relations with anybody is through [the Integrative Unit], but if we were brought in earlier, I think it would help the researchers.
>
> As a source of market information, the field sales force is so spread out and working on so many things they can't give us good information. They should be a very good source about the primary need, but they aren't. Salesmen come back and tell us about the need in such broad generalities that it sounds like they are talking about superproducts. They say they would like a clear plastic that can take any color and be completely heat resistant and easy to shape, etc. Any idiot knows we all want that, but what we need is more specific knowledge coming from qualified persons, such as the salesmen.
>
> We lack coordination in getting the new product from the laboratory to the field. It sometimes takes two years to get the field people to push a new product. One thing we have been trying to do is to get the technical person out into the field so that he can push his product himself.

In summary, neither of these low-performing organizations met the demands of the environment for high differentiation and integration so well as either the medium- or the high-performing ones. The medium performers, while they managed to achieve either the required differentiation or the required integration, failed to achieve both. The high performers did most nearly meet both the requirement for highly differentiated units and the necessity for these units to work well together. There appears to have been a close relationship between the extent to which these organizations met the environmental requirements for differentiation and integration and their ability to deal effectively with that environment.

But this finding still leaves us with a curious contradiction. If, as we have found, differentiation and integration work at cross purposes within each organization, how can two organizations achieve high degrees of both? The best approach to explaining this apparent paradox becomes evident if we consider how organizations might go about achieving both of these states. If organizations have groups of highly differentiated managers who are able to work together effectively, these managers must have strong capacities to deal with interdepartmental conflicts. A high degree of differentiation implies that managers will view problems differently and that conflicts will inevitably arise about how best to proceed. Effective integration, however, means that these conflicts must be resolved to the approximate satisfaction of all parties and to the general good of the enterprise. This provides an important clue to how two of these organizations met the environmental requirements for high differentiation and high integration. These two organizations differed from the others in the procedures and practices used to reach interdepartmental decisions and to resolve conflict. We now want to explore these differences.

CHAPTER III

Resolving Interdepartmental Conflict

IN CONSIDERING how these six organizations vary in their capacities to resolve interdepartmental conflict, we must find answers to two closely related questions. First, *what factors within the organizational system determine whether managers deal effectively with interdepartmental conflict?* Recent studies of managerial behavior in organizations, including our own earlier work, suggest several factors that can be related to how effectively managers handle interdepartmental conflict.[1] We mention these earlier findings not to explore them here, but because they raise the second, and more important, question: *Are these determinants of effective conflict resolution more likely to be present in the organizations that are both highly differentiated and highly integrated than in the other four organizations?*

If these determinants are more typical of the successful organizations, we have at least a partial answer to the riddle of how two organizations achieved both the differentiation and the integration necessary to deal with the dynamic and diverse market, technical, and economic conditions confronting them. To carry out a systematic investigation of these two questions, instead of considering all the possible sources of conflict among departments, we limited our inquiry to the single most crucial competitive issue of the plastics industry—product and process innovation. We assumed that the handling of conflicts around this key issue would, in itself, be an important determinant of effectiveness and, beyond that, would serve as a good indicator of how conflicts around lesser issues were resolved.

As a background for examining the findings on conflict resolution, it is necessary to examine two related points. First, given the nature of the different parts of the environment, who in these organizations had the required knowledge to help resolve interdepartmental conflicts involving product and process innovation? Second, what managers in these organizations had been assigned the responsibility for resolving these conflicts?

The Requirements for Interdepartmental Conflict Resolution

The varying degrees of uncertainty in the three major parts of the organization's environment (market factors, techno-economic factors, and scientific factors) meant that personnel at different levels within each department dealing with these different factors had the knowledge necessary to help make joint decisions on innovation issues with other departments. We found in interviewing top executives about environmental conditions that, given the relative stability of techno-economic factors, a high-level production manager (*e.g.*, plant manager) knew enough to consider joint departmental decisions from the production viewpoint. On the other hand, the great uncertainty about scientific facts and their application meant that only the scientists working on particular problems or their immediate supervisors (*e.g.*, group leaders) had the detailed knowledge necessary to contribute to the resolution of problems with other departments. The market knowledge needed for making joint decisions was less certain than the techno-economic information, but more certain than the scientific knowledge. This meant that managers nearer the middle of the sales department hierarchy (*e.g.*, product sales managers or product marketing managers) were the ones with the necessary knowledge to attempt integration and to work at resolving interdepartmental conflict.

In addition to these representatives of the basic functional departments, each organization, as we indicated in the previous chapter, had established a department that had as one of its major responsibilities the integration of the activities of the basic departments. While this was the assigned responsibility, our observations led us quickly to conclude that the primary activity in bringing about integration was helping to resolve conflicts among the different functional points of view. Given the uncertainty about both the market and the scientific phases, and given the fact that the effective integrator had to know something about all facets of the environment, the top managers we interviewed indicated that the integrators generally had to be at the lower management levels in order to have the knowledge required to carry out the detailed integrating activity that the environment demanded.

This is not to say that the top echelons of management in the research, sales, and integrating departments were not supposed to be involved in integration. Rather, the uncertainty and complexity of the market and scientific aspects of the environment limited the extent to which these upper managers could get involved in detailed decision making. Their job was to provide integration at a broader level of policies and strategies.

In all six organizations we found that the managers at the required level in each department had been assigned the formal responsibility for integrating their departments' efforts with those of other units around product and process changes. If this had not been so, we might have immediately had one obvious clue why some organizations were more effective than others at resolving conflict to achieve both differentiation and integration. But while the assignment of responsibility met the requirements of the environment, what actually went on in some organizations differed from what was formally designated. This is a point we will pursue in some detail. Now we wish simply to state that formal responsibility for integration (and thus for the resolution of conflict) was assigned at the required level within each functional department and the coordinating departments.

The men to whom these responsibilities were assigned were plant managers from production, scientists and group leaders from research, and product sales and marketing managers from sales, as well as the assigned integrators. In four organizations a rather complex network of cross-functional teams or committees had been formally established to provide a setting for these managers to carry out joint decision making around a particular product or group of products. In the other two, even though no formal teams had been set up, there were frequent meetings of integrators and functional managers, either in groups or in pairs, to achieve integrated efforts.

While all these managers were supposed to be involved to some extent in resolving interdepartmental conflict, we came to recognize that the integrators were the most active. This was a major part of their assigned work. They were presumably on the payroll largely because of their ability to do it, while the various functional managers were paid chiefly for their specialized knowledge and skills. As we thought about the factors that could be related to conflict resolution in these organizations, we realized that one important set of determinants would revolve around the behavior of the integrators, while another might be related to the behavior of both the integrators and the functional specialists. In examining the six determinants of conflict resolution, we shall focus first on three factors that influenced the integrators as they attempted to facilitate the resolution of interdepartmental disputes and then upon the factors that seemed to be related to the behavior of all the managers involved.

Determinants of Effective Conflict Resolution

In discussing the factors that determined how these organizations dealt with conflict, we will often use the terms "conflict resolution" and "joint decision making" interchangeably. From our observations, both terms refer to different stages of a single process. While the objective of the managers was to reach interdepartmental decisions that would constitute commitments from each department to carry out part of a coordinated action plan, reaching these decisions invariably involved the resolution of conflicts among departments. The desired end was a joint decision, but the means to achieve it was the resolution of conflict.

We should also emphasize that although prior research into human behavior in organizations had suggested some possible determinants of how effectively conflict was handled in an organization, we had no prior evidence that these determinants might meaningfully discriminate between the more effective and the less effective organizations in this environment. In fact, when we started this study, we were predicting that the only real discrimination would be in the total configuration of practices and methods of handling interdepartmental conflict, and not in any one or two characteristics. In reporting our findings, however, we shall separately review each individual determinant and then compare all six organizations in terms of the pattern of the six factors that we found did distinguish the high-performing organizations from the low-performing ones in this industry.

1) *Intermediate Position of Integrators*

The first factor that we predicted might enable the integrators to be more effective in their jobs had to do with their own orientations and the structure of their departments. If the integrators had developed goal, time, and interpersonal orientations equidistant among those of the managers in the various departments they were linking, we predicted they would be more effective in resolving conflicting viewpoints than if they tended clearly toward one orientation.[2] Similarly, we expected greatest effectiveness if the structure of the integrating department was intermediate between that of the highly structured production unit and the least structured research laboratory. In essence, if the integrators thought and behaved in ways that were not too dissimilar to those of each of the departments with which they had to work, they would be better able to communicate with all these departments and thus to help solve interdepartmental disagreements. If the integrators adopted points of view and behavioral patterns that were too similar to those of any one functional department, the others would see them as understanding only the problems of that particular unit and not as responsive to the problems of other units.

Our findings seem generally to support these predictions (Figure III–1). In one of the two high-performing organizations (high performer A), the integrating department had an intermediate structure, and its members had developed approximately equidistant time, goal, and interpersonal orientations. In one of the two low-performing organizations (low

Figure III–1

POSITION OF INTEGRATING DEPARTMENT IN STRUCTURE AND ORIENTATION [a]

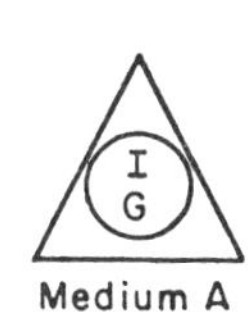

[a] Letters in center circle indicate the attributes **G**oal, **T**ime and **I**nterpersonal orientations and **S**tructure in which the integrating department in each organization was intermediate among the basic functional units. See Methodological Appendix for discussion of how intermediate position was computed.

performer A), the integrators were intermediate in only one attribute (interpersonal orientation). In the other four organizations, the integrators were intermediate among the functional departments in only two of the four attributes we measured.

In interviews managers in the medium- and low-performing organizations indicated that the lack of balanced thinking on the part of integrators limited their ability to facilitate interdepartmental decision making. They expressed most concern about a lack of balance in either the time or the goal orientation of the integrators. For example, the functional managers in medium-performing organization A and in the two low-performing organizations, where the integrators were not intermediate in time orientation, complained frequently that the integrators were too preoccupied with current problems to be helpful in dealing with longer-range matters. For example, below are some typical comments from managers:

> I am no integrator, but I can see that one of our troubles is that the integrators are so tied up in day-to-day detail that they can't look to the future. They are still concerned with [this year's] materials when they should be concerned with [next year's] markets.
>
> * * * * *
>
> We get lots of reports from [the integrators] and we talk to them frequently. The trouble is that all they present to us [in research] are short-term needs. They aren't the long-range things in which we are interested.
>
> * * * * *
>
> [The integrators] only find out about problems when they find out somebody has quit buying our material and is buying somebody else's, and this keeps you on the defensive. A lot of our work is catch-up work. We would like more future-oriented work from them.

Similarly, functional managers in medium-performing organization B and low-performing organization A frequently complained about the integrators' lack of balance in goal orientation:

> Our relations with [the integrators] are good, but not as good as with research. The [integrating] people are not as cost conscious as the laboratory people. They are concerned with the customer.
>
> * * * * *
>
> The [integrator] is under a lot of pressure to work with salesmen on existing products in our product line. What the [integrator] should be and often tries to act like is a liaison person, but in reality he is not. He is too concerned with sales problems.
>
> * * * * *
>
> What's lacking is that the [integrators] are so busy that they continually postpone working with research. They work closely with applied research with minor modifications, but the contact with basic research is minimal.

The fact that the managers in these four organizations expressed most concern with the lack of balance in time and goal orientation does not mean, in our view, that interpersonal orientation and structure were not equally important. Instead it suggests that the imbalances they mentioned were more obvious to them. The integrating departments in these organizations were also not intermediate in structure or interpersonal orientation of members. This was probably a source of misunderstanding in many cases, but the managers were not manifestly aware of the difficulties caused by these differences.

One example of how the structure of the integrating department had an impact on the effectiveness of the integrators occurred in the case of a production manager in low-performing organization A. This manager complained that the integrators were constantly coming to him about matters that should be handled several levels higher up the production hierarchy. From the point of view of the integrators, accustomed in their own department to emphasis on direct contact among managers at lower levels, this seemed appropriate. But to this production manager, in a highly structured department where rules and procedures specified who did what, the integrators' behavior was an irritant, which reduced his effectiveness in working with the plant manager on matters that the latter could handle. The plant manager did not realize that it was the difference in structure that created his problem, but to us this seemed to be the case.

In this discussion we have made no mention of high-performing organization B. Even though its integrators were intermediate in only two of the attributes, all the evidence available from both questionnaires and interviews indicated that they were effective in bringing about integration among the basic departments. One possible explanation could be that the integrators were intermediate in the two attributes (time and goal orientations) of which other managers were most aware. There appears, however, to have been another, more important, factor, which we shall come to shortly.

2) *Influence of Integrators*

The second factor that we predicted might help to determine how effective the integrators would be in resolving interdepartmental conflict was how much influence they appeared to have in making the relevant joint decisions.* We predicted that the integrators would need to be seen by others in their organization as having an important voice, in fact *the* most important voice, in decisions. Our reasoning was that, since none of the basic functions was uniquely critical for achieving the key goal of innovation, it would follow that integration was in itself the most problematic job. High

* That this factor might be important was suggested by the work of John A. Seiler, "Diagnosing Interdepartmental Conflict," *Harvard Business Review,* September–October 1963. The methods used to measure influence are described in the Methodological Appendix.

weight given to integrators would lend prestige to their task and increase the likelihood that it would be achieved. As it turned out, the integrators in all six organizations were seen by their organizational colleagues as having high influence relative to the managers in the functional departments. In all six organizations, the integrating department was seen as either the first or second most influential in the organization on the critical issues of innovation (Table III–1). Therefore, even though this high influence may have been an important factor in the integrators' effectiveness, we have not explored it further in this comparison, because it did not discriminate among these organizations.

Table III–1

INFLUENCE OF INTEGRATORS RANKED RELATIVE TO THE BASIC FUNCTIONAL DEPARTMENTS

Organization	*Rank*
High performer A	1
High performer B	2
Medium performer A	1.5 [a]
Medium performer B	1
Low performer A	2
Low performer B	1

[a] Tied for first ranking.

While the managers' assessment of the integrators' relative influence did not vary among these organizations, we were aware, as a second aspect of this issue, that the basis upon which this influence was built could very well differ. Sociologists have pointed out that a person's influence within an organization can stem from a number of factors—the formal position he occupies, his expertise or competence, or perhaps his age or length of service. More specifically, we were interested in the notion suggested by earlier writers that in a situation where expert judgments were required, influence over decisions based on a person's position alone would not lead to effective problem resolution.[3] In this industry, where joint decisions had to combine the expert judgment and knowledge of production, marketing, and scientific specialists, it appeared likely that integrators would not be helpful in resolving conflicts unless the specialists whose efforts they were tying together saw them as competent and knowledgeable.

As the functional managers spoke about their relationships with the integrators, it became apparent that there were, in fact, important differences in the reasons for the integrators' high influence in the two high-performing organizations as compared with the others. In the former, the integrators were seen as being highly competent, and the managers related this expertise to their influence over decisions. Sample quotations from the managers in these two organizations illustrate this point:

> The [integrator] has a powerful job if he can get the people to work for him. A good man in that job has everybody's ear open to him. A good [integrator] has to be thoroughly oriented toward his market or to his process. Whichever area he is working in, he has to be able to make good value judgments.

* * * * *

> The [integrators] are the kingpins. They have a good feel for our [Research's] ability, and they know the needs of the market. They will work back and forth with us and the others.

* * * * *

> The [integrator] is on the border of research, so we work closely together. The [integrators] are just a step away from the customer, so when I make a change in a material, I let them know because they may have a customer who can use it. The good thing about our situation is that the [integrators] are close enough to sales to know what they are doing and close enough to research to know what we are doing.

What the managers in these two organizations were saying was that the integrators were influential because managers valued their knowledge and their expertise. When conflicts arose, all concerned listened to the integrators because they respected their competence. The integrators in the other four organizations, who also had great influence over decisions, were listened to for quite different reasons. As the typical comments below indicate, these integrators had an important voice not so much because of the authority of knowledge, but more because of the formal authority given them by top management and their close reporting relationship to top managers:

> A good [integrator] is a guy with a red hot bayonet. He doesn't take "no" for an answer on anything. He is also in an enviable position, since he reports to the general manager and finds very little opposition to what he wants to do.
>
> * * * * *
>
> Nobody wants to pull the wool over the integrator's eyes, since he reports to the general manager. To do so would be disastrous for the individual. I don't think anybody could be an [integrator] and have many friends. You have to be too aggressive.

The managers interviewed in these four organizations also emphasized another related point. They did not see the integrators as being particularly knowledgeable or expert:

> For a man to move into [the integrating unit] should be a big thing. But it isn't now. My guys [in research] can say I know more than that guy [in integrating]. People compare their skills with those of the integrators, and often the comparison is not favorable.
>
> * * * * *
>
> The integrator is supposed to know the field, and he may think our product isn't any good. This is fine if you have confidence in him, but we have had a bad experience with some of them. As the knowledge of chemistry grows, the integra-

> tors' knowledge of the market must grow. I guess I would appraise the situation this way: just because the integrators have had twenty years' experience doesn't mean they have twenty years of knowledge.

What these data suggest is not that authority stemming from the hierarchy is unnecessary, but rather that in solving complex problems it is possible for this type of influence to be inconsistent with the influence derived from knowledge about the problems to be solved. In the high-performing organizations the two kinds of authority coincided and reinforced each other. The integrators had the assigned authority to help resolve conflict, and their colleagues felt that they knew enough to carry out this assignment. In the other four organizations the integrators had the same assigned authority, but since they had less influence derived from knowledge, the two influence systems came into conflict. The functional managers felt that, in terms of knowledge and expertise, they themselves had more to contribute to decision making, and they resented the positional influence of the integrators. While they recognized that the integrators were influential, they did not always feel that their decisions were consistent with the facts of the situations. This seemed to lead to less effective resolution of conflict and, as important, to less motivation on the part of functional specialists to follow up on decisions reached.

We have no firm explanation of why the integrators in the high-performing organizations were seen as influential because of their competence while those in the other organizations were not. One possible explanation is that only in these two organizations were the integrators balanced in both time and goal orientations. This balance in the orientations of which the various specialists were most aware may have led the specialists to see the integrators as competent. But a second, and perhaps more important, reason why managers thought the integrators' influence was derived from competence was that higher management did not pressure the integrators to use the authority of their position to force decisions in directions that the higher manager preferred.

Before we examine the third factor in the integrators' effectiveness in resolving conflict, we should emphasize that in discussing the basis of influence we have been focusing on the extent to which these integrators were *seen* as competent and knowledgeable, and not on the extent to which they actually were. While we made no systematic attempt to compare the actual knowledge of integrators in these six organizations, all the readily available information, such as education and experience, would suggest that all the integrators were in fact quite similar in their actual knowledge about the technical and market problems facing them.

Reward System for Integrators

The extent to which integrators felt they were evaluated and rewarded in accordance with the overall performance of their product group was the third factor that we expected could be related to their effectiveness in resolving conflicts.[4] If the integrators felt that their superiors were evaluating them on the basis of the profitability of their product group or on a similar overall standard of performance, they would be more motivated to work at achieving integration. Since, as we have pointed out, all of these integrators had some other responsibility in addition to integration (*e.g.*, supervision of development activity or marketing planning), when they felt that their own individual performance or the performance of their subordinates was the most important factor in their evaluation, they might devote less effort to purely integrative activities.

Information about the importance of different bases of reward in each organization was collected as part of the questionnaire.* The integrators in the two low-performing organizations reported that, for them, the total product-group performance was significantly less important as a basis of evaluation than did their counterparts in high-performing organization B and medium-performing organization A (Table III–2). Interestingly enough, the integrators in the two low-performing organizations also reported that their

Table III–2

INTEGRATORS' PERCEIVED BASIS FOR REWARD [a]

Organization	*Product group performance*
High performer A	2.0
High performer B	1.8 [b]
Medium performer A	1.1 [b]
Medium performer B	2.0
Low performer A	2.5 [b]
Low performer B	3.0 [b]

[a] 1 = most important basis
2 = second most important basis
3 = third most important basis

[b] The two low-performing organizations were significantly different from high-performing organization B and medium-performing organization A at .01 level (Orthogonal comparison).

own individual performance was a significantly more important factor in their superiors' evaluation of them than did the integrators in the other organizations. These data suggest that the integrators in the organizations with the most problems in achieving the required states of both differentiation and integration were less concerned with the superordinate goals of total product performance than were their counterparts in the other organizations. The integrators in high-performing organization B and medium-performing

* For details of this part of the questionnaire, see the Methodological Appendix.

organization A were probably most concerned about these goals, since they indicated that this was a significantly more important factor in their evaluations, while the integrators in high-performing organization A and in medium-performing organization B were concerned to a moderate extent with these superordinate product objectives.

Integrators in these four organizations were apparently getting either explicit or implicit signals from their superiors that a relatively important part of their job was to contribute to the overall performance of the products for which they were responsible. In the low-performing organizations this particular incentive to work at the resolution of conflict was not so noticeable, which may have contributed to the difficulties these organizations encountered in achieving the required state of differentiation and integration.

So far we have reported on three determinants of effective conflict resolution which are especially related to the behavior of the integrators—intermediate orientation and structure; high influence based on competence and knowledge; and the presence of incentives for conflict resolution. Each of these three factors has to some extent discriminated among the more effective and the less effective organizations. As we have already indicated, our interest is not in whether any one or two factors distinguished the high-performing organizations from the lower ones, but rather in the total configuration of the several factors that seem to determine how effectively conflict is handled in an organization. Before we can examine the overall pattern of these determinants of effectiveness, we need to consider three other factors that had an impact on the decision-making behavior of the functional specialists as well as that of the integrators.

4) *Total Level of Influence*

The first of these factors is the amount of influence managers in all the departments felt that they had over decisions.

If the various functional specialists and the integrators all felt that their own departments had great influence in the decision-making process, they would all be likely to feel that their point of view was considered as decisions were reached. As a result, they would be likely to feel less hostile toward other departments even when a particular decision did not seem ideal from their point of view.[5] Essentially, we were predicting that the process of resolving conflicts would produce less heat and more light when all managers felt that their own departments had an important voice in interdepartmental decisions.

Managers in the two low-performing organizations did report significantly lower influence throughout all departments than did those in the other four organizations (Table III–3).* The managers in these two organizations felt that they had significantly less say in decisions than did managers in the other four organizations. Thus the total amount of influence all managers attributed to themselves was different in the two organizations that had the most difficulty in

Table III–3

TOTAL INFLUENCE IN THE SIX ORGANIZATIONS

Organization	*Mean influence scores* [a]
High performer A	3.6
High performer B	3.6
Medium performer A	3.5
Medium performer B	3.6
Low performer A	2.6 [b]
Low performer B	3.1 [b]

[a] Possible scores ranged from 1. –little or no influence, to 5. –a very great deal of influence.

[b] Low-performing organizations were significantly different from other organizations at .05 level (Orthogonal comparison).

* The method for measuring influence was the same as that used for measuring the influence of integrators (see the Methodological Appendix).

achieving the required states of differentiation and integration from that in the other organizations.

This finding, however, needs some clarification. It is clear that influence in an organization is not a fixed commodity passed down from the top of the organization, as it is sometimes assumed to be. Instead, our findings suggest, as other researchers have found, that influence can expand or contract.[6] In four of these organizations the typical manager felt, because of his knowledge and the positional influence delegated to him, that he had considerable influence over decisions reached. As he and his colleagues exercised this felt influence through interaction and discussion, their total amount of influence in the organization did grow. There are two principal explanations of the link between high total influence and organizational effectiveness. One, which was our initial theory described above, is that the connecting link is improved motivation: If all the managers feel that their views are given sufficient attention, they are less likely to be dissatisfied with decisions or to feel antagonistic toward other departments, and better motivated to act vigorously on their part of the agreed-on plan of action. The alternative explanation is that, since the knowledge and judgment of all the departmental specialists were relevant to sound decisions, giving them all considerable influence simply improved the quality of the resulting decision. By this we mean that the decision took more realistic account of environmental facts and thereby led directly to better performance. The data we have examined above do not help us test out these two explanations, since either or both could be at work. We will come back to this question with additional data in Chapter V.

5) *Influence Centered at Required Level*

Another way in which influence seemed to be tied to the process of resolving interdepartmental disputes was in the requirement that within each functional department influence be concentrated at the managerial level where knowledge to make decisions was available. When the managers who had the requisite knowledge also felt that they had the necessary influence, they would be more effective at resolving conflicts.[7] If the managers who had the knowledge to make decisions did not feel that they had enough influence, perhaps because higher managers did not behave in accordance with assigned authority, the decision-making process would suffer. The managers who were making decisions would not have the knowledge to do so, and the managers who had the knowledge would feel left out. Once again a dual explanation is possible. The decisions made might not be based on the available knowledge, or the knowledgeable managers might not be motivated to implement decisions that they felt were not based on the facts of the situation.

We have already indicated that the knowledge required to make decisions is available at different levels in each of the functional departments (at the lower levels in research, at the middle level in sales, and at the upper levels in production).* In both low-performing organizations we found that influence was not concentrated at the required level in two of the departments (Table III–4). The managers at the upper level of the applied research department in low-performing organization B had too large a voice in decisions, while the lower-level managers in the production department seemed to have too much influence relative to their knowledge. In low-performing organization A the influence in both research departments was concentrated at too high a level.

In high-performing organization B and the medium-

* The required knowledge was available at the lower levels of the integrating departments. However, since in all six organizations influence was greatest at these levels, which was consistent with the requirements, the integrating departments were excluded in this part of the analysis. The method used to measure influence at different levels within each department is described in the Methodological Appendix.

Table III–4

CONGRUENCE OF LEVEL OF INFLUENCE AND REQUISITE KNOWLEDGE

Organization	*Congruent in 4 departments*	*Congruent in 3 departments*	*Congruent in 2 departments*
High performer A	X		
High performer B		X	
Medium performer A		X	
Medium performer B		X	
Low performer A			X
Low performer B			X

performing organizations, each had one department where the influence was concentrated at a level that was not consistent with the availability of the required knowledge. In high performer B influence was centered too low in the production department hierarchy. The concentration of influence in the applied research laboratory in medium performer A was at too high a level, while in the sales department of medium performer B it was too low.

High-performing organization A was the only one in which the concentration of influence fit the required knowledge in all four departments. This organization, with both high differentiation and integration, met this condition completely. While high-performing B and the two medium performers met it partially, the two low-performing organizations met it to the least extent of any of the six.

Modes of Conflict Resolution

The final factor that distinguished the conflict-resolving effectiveness of the two high-performing organizations from that of the others was perhaps the most obvious one—differences in the mode of behavior typically used to deal with conflict. When faced with an interdepartmental conflict, managers can and do respond in several ways.[8] We anticipated that the most effective way for them to handle conflict would be through a problem-solving or confrontation approach. If the managers involved openly exchange information about the facts of the situation as they see them, and their feelings about these facts, and work through their differences, the probability of reaching a solution that is optimal for the whole organization should be greatest.

To clarify what we mean by confrontation we can cite some comments from managers:

> Our problems get thrashed out in our committees. We work them over until everybody agrees this is the best effort you can make. We may decide this isn't good enough. Then we decide to ask for more plant, more people, etc. We all have to be realistic and take a modification sometimes and say this is the best we can do.
>
> * * * * *
>
> In recent meetings we have had a thrashing around about manpower needs. At first we didn't have much agreement, but we kept thrashing around and finally agreed on what was the best we could do.

Not only did we anticipate that confrontation would be the most effective method of resolving conflict, but we also learned from questionnaire data that in the judgment of the managers in these six organizations this was the ideal way in which conflict should be resolved. Confronting conflict, however, requires a great deal of emotional and intellectual energy as well as a high degree of interpersonal skill. Consequently, other ways of handling conflict are also prevalent in some organizations.

Instead of confronting conflict, the managers involved can elect to split the difference and reach a compromise satisfactory to the various parties. The difficulty is that a compromise resolution often does not meet the needs of the whole organization. Another mode of handling conflict is to pour oil on the troubled waters and smooth them over. Unfortunately, when problems are smoothed over, they usually do not seem to solve themselves; in fact, they often fester and become worse when no action is taken. In these cases the managers seem to be saying, "We are all good friends, and we shouldn't let this problem disrupt our friendship, so we will just ignore it and hope it works itself out." Another comment from a manager in one of these organizations may help to illustrate what is meant by the smoothing of conflict:

> I thought I went to real lengths in our group to cause conflict. I said what I thought in the meeting, but it didn't bother anybody. I guess I should have been harsher. I could have said I won't do it unless you do it my way. If I had done this, they couldn't have backed away, but I guess I didn't have the guts to do it. I guess my reaction was—well I made a fool of myself in the meeting and nothing happened so I'll sit back and feel comfortable. I guess I didn't pound the bushes hard enough.
>
> The relations are wonderful. We are happy—we are friendly and happy as larks. We chew each other out and are happy and go about our business. I've never run into more cooperative people. I think they think we are cooperative also, but nothing happens.

Finally, interdepartmental conflicts can be handled by one side or the other trying to use the power of knowledge or position to force a solution that is satisfactory from only one point of view. One way to do this is by asking a superior to help get a unilateral resolution. As one manager described it:

> If I want something very badly and I am confronted by a roadblock, I go to top management to get the decision made. If the research managers are willing to go ahead [my way], there is no problem. If there is a conflict, then I take the decision to somebody higher up.

Similarly, a resolution can be reached by two managers forming a solid front that can be used to force others to accept their desires. One manager outlined how this worked:

> I really think our team doesn't act as a team. We have lots of meetings that consist of two people on a team. It is not unusual to get a call from the production representative asking me to come over. We would sit down together and he would explain that he had a problem that wasn't being attacked properly. He would say that the [integrator] wasn't supporting him. This should be handled in our team. We find small groups getting together and discussing things, because they think the other two members won't agree. Well, this isn't acting as a team. It's our weak spot.

The obvious disadvantage of the forcing mode is that some relevant knowledge is often ignored, and the other parties to the dispute, feeling that their wishes have not been taken into account, are not likely to be motivated to carry out the decisions reached.

Our analysis of questionnaire data identified three distinct modes of actually handling conflict in these six organizations: *Confrontation,* or problem-solving; *smoothing-over* differences; and *forcing* decisions.* The analysis did not distinguish compromise as a separate mode. This does not imply that it was not used in any of these organizations, but only that our measurement methods did not enable us to identify it as a distinct means of handling conflict.

The questionnaire data indicated important differences among these organizations in the typical modes managers used to handle interdepartmental conflict. While the managers in each organization indicated that confrontation was their most usual mode, there were significant differences in how typical this behavior was in the various organizations (Table III–5) . In the high-performing organizations confrontation was apparently used to a significantly greater degree than in the other four organizations. The two organizations

* The method used to gather data about typical modes of resolving conflict and the procedure used to analyze these data are described in the Methodological Appendix.

Table III–5

MODES OF CONFLICT RESOLUTION

Organization	*Confrontation*	*Smoothing*	*Forcing*
High performer A	13.0 [a]	8.9	9.5
High performer B	13.1 [a]	9.3	9.5
Medium performer A	12.4 [a]	9.0	9.0 [b]
Medium performer B	12.0 [a]	9.8 [b]	9.7
Low performer A	11.7 [a]	9.0	9.8
Low performer B	11.8 [a]	9.8 [b]	8.5 [b]

Higher score indicates *more* typical behavior.

[a] Pairs of organizations (high performers, medium performers and low performers) significantly different from other organizations at .01 (Orthogonal comparison).
[b] Significantly different from other organizations at .01 (Orthogonal comparison).

with the high states of differentiation and integration required to deal effectively with their environment had managers who relied on confrontation significantly more than did the managers in either the medium- or low-performing organizations. Similarly, managers in the medium-performing companies were using confrontation to a significantly greater extent than were those in the low-performing organizations.

These data also indicated that there were differences among organizations in the extent to which managers smoothed over conflicts and disputes. The managers in medium-performing organization B and low-performing organization B were evidently significantly more prone to smooth over conflicts than were managers in the other organizations. The managers in these two companies were not only relying less on confrontation than were the managers in the two most effective organizations, but were also doing significantly more smoothing than the rest of the managers.

The managers in low-performing organization B also reported that they were doing significantly less forcing than the managers in the other organizations. Thus the managers in the organization that was both least differentiated and least integrated were emphasizing the smoothing over of differences and the maintenance of friendly relations to the extent that it seriously impaired their ability to reach interdepartmental decisions. A less serious but similar situation seemed to occur in medium-performing organization A, where the managers were also doing significantly less forcing than in the other organizations. All of these data taken together suggest that while heavy reliance on confrontation is essential, it may also be important in organizations in this environment to have a back-up mode that relies on some forcing and a relative absence of smoothing. Perhaps the presence of some forcing means that the managers will at least reach some decision instead of avoiding problems.

In summary, as we had predicted, reliance on confrontation seems to lead to effective conflict resolution and to the desired states of differentiation and integration. The high-performing organizations met this determinant most clearly. The medium-performing ones to a moderate extent; and the low performers the least.

Comparative Pattern of Determinants of Effective Conflict Resolution

So far we have compared these organizations separately in each determinant of effective conflict resolution, both those that affected the behavior of the integrators only and those that had an impact on all managers. As we have done so, certain differences between the highly effective and the less successful organizations have become noticeable. As we stated earlier, however, we were not interested in the way any one of these determinants discriminated among these organizations, but in comparing the overall pattern of these determinants. To make this comparison we have indicated which organizations met each of these determinants to a high degree, to a medium extent, and to only a low degree (Table III–6) . In comparing these patterns, we should emphasize that there is at present no adequate theory or empirical evidence to guide us in judging the relative impact of each of these conditions. Certainly their impact is not simply additive. Similarly, while these conditions, like other determinants of behavior in organizational systems, are interrelated, we do not yet have sufficient knowledge to describe the precise relationship among them. For example, it is clear that in the high-performing organizations, where the integrators were seen as having influence based on their competence, the managers relied more on confrontation to reach decisions than did managers in the other four organizations. One can speculate that there is some relationship between these two facts. Perhaps in the organizations where the integrator's influence was based more on positional authority, the integrators attempted to force decisions, which led other managers to respond with behavior other than confrontation.

While we could speculate on this and other relationships, as well as on which conditions had the greatest effect on decision-making behavior, it does not seem fruitful to do so with the data available. Rather, we have summarized these data so that the total configuration of determinants can be visually compared among organizations. From this comparison it is clear that in the two high-performing organizations there were more conditions that would lead to effective resolution of interdepartmental conflict. This overall finding provides an important explanation of how these two organizations managed to attain both high differentiation and integration, even though the two states were essentially antagonistic. Managers in these organizations were apparently able to deal effectively with conflict. In spite of their widely different ways of thinking and behavior patterns which enabled

Table III–6

SUMMARY OF PARTIAL DETERMINANTS OF EFFECTIVE CONFLICT RESOLUTION RELATIVE TO DIFFERENTIATION, INTEGRATION, AND PERFORMANCE

Organization	*Intermediate position of integrative organization* [a]	*Influence of integrators derived from technical competence* [a]	*Integrators perceive rewards related to total performance* [a]	*High influence throughout the organization* [a]	*Influence centered at requisite level* [a]	*Modes of conflict resolution* [a]	*Degree of differentiation*	*Degree of integration*	*Organization System performance*
High A	High	High	Medium	High	High	High	High (9.4)	High (5.7)	High
High B	Medium	High	High	High	Medium	High	High (8.7)	High (5.6)	High
Medium A	Medium	Low	High	High	Medium	Medium	Low (7.5)	High (5.3)	Medium
Medium B	Medium	Low	Medium	High	Medium	Medium	High (9.0)	Low (5.1)	Medium
Low A	Low	Low	Low	Low	Low	Low	High (9.0)	Low (4.9)	Low
Low B	Medium	Low	Low	Low	Low	Low	Low (6.3)	Low (4.7)	Low

[a] High, medium, or low indicates relative extent to which each organization met each determinant.

them to perform their specialized tasks, managers were able to reach decisions that provided effective collaboration.

The medium-performing organizations met these conditions for effective conflict resolution to a lesser extent than did the high-performing organizations but to a greater extent than the two low-performing organizations. These two medium performers had reached an apparent trade-off between the required differentiation and the necessary integration. Each organization had achieved one state, but not the other. Since medium-performing organization A was not highly differentiated, its managers were still able to achieve effective collaboration. Medium-performing organization B, on the other hand, had achieved the required differentiation, but its managers were not able to work across their different orientations and structures to resolve conflicts effectively and reach well-integrated decisions. All of this suggests that the inability of these managers to resolve conflict effectively contributed to their medium level of performance.

The situation in low-performing organization A appears similiar to that in medium-performing organization B, only perhaps worse. While this low performer had achieved the required differentiation, it met fewer of the conditions determining effective conflict resolution than did medium-performing organization B. Consequently, its managers were even less effective at reaching joint decisions, and integration was even lower. The other low performer (low-performing organization B) also was very low in meeting all of these determinants of effective conflict resolution. As a result, it had achieved neither the differentiation nor the integration required in this environment. Even though its managers in various departments were closer together in their orientations and organizational practices than those in other organizations, their low effectiveness in handling conflict meant that they could not achieve the integration necessary to deal effectively with external conditions.

While these determinants of effective conflict resolution offer a good explanation of how organizations achieved both differentiation and integration, we need only compare the two low performers to see that it is probably not the entire explanation. Low performer A, which met these conditions to about the same extent as low performer B, did manage to achieve both better integration and higher differentiation. We have no further explanation of why this happened. We mention it only to illustrate that our knowledge is not complete. In spite of this limitation, however, it seems safe to conclude that organizations that are effective in a dynamic and diverse industry will have to meet these and perhaps other conditions that lead to effective resolution of interdepartmental conflict. In this way their managers will be able to maintain their highly specialized points of view and behavior patterns, while they work together to achieve a joint effort that is productive for the organization's total objectives. In essence, the evidence indicates that high differentiation plus effective conflict resolution leads to high integration, and these are the overall conditions organizations must attain to be effective in this type of environment.

In considering these findings and their implications, the reader should be aware of an important fact. The use of these concepts of differentiation and integration, as well as those dealing with the determinants of effective conflict resolution, has enabled us to better understand the functioning of large organizations in a particular kind of environment. But other environments place other demands on organizations and their members. Most obviously, less differentiation may be required of organizations in other types of industry. Equally important, but perhaps less obvious, some of these determinants for effective conflict resolution may be different in other industries. For example, the level at which the influence to resolve conflict needs to be located may vary with the uncertainty of knowledge about the environment. In more certain environments influence may be centered at higher levels in all departments without loss of effectiveness. Similarly, in other industries, where particular aspects of the environment are of strategic significance, a unit other than the integrating department may be required to be most influential.

While this description of how effective organizations in other industries need to be different from the high-performing plastics organizations is only suggestive of the full extent of possible difference, the essential point is that we have found that such differences do exist. Our purpose in the next chapters is to investigate, in detail, how high-performing organizations in other environments differed from these firms. As we turn to this question we should issue an important *caveat* to the reader. What we have described so far is relevant only to understand the functioning of organizations under one set of technological, scientific, and economic conditions. Only after we have examined the requirements in other types of industrial setting will we be in a position to consider the implications of this study for the broader practice of administration and for organization theory. We ask the reader to resist the temptation to reach more general conclusions until we have described the second major part of this study.

CHAPTER IV

Environmental Demands and Organizational States

OUR PRIMARY objective in the next three chapters is to see how effective organizations in other industrial environments differ from those in the plastics industry in states of differentiation and integration and in procedures for resolving conflict. A second important aim is to learn whether the findings about the differences between high-performing and low-performing organizations in the plastics industry are confirmed by the comparison of organizations with similar performance records in other industries. We shall now examine four additional organizations: a high-performing and a low-performing one (by conventional economic criteria) in each of two industrial environments. We shall look first at the states of differentiation and integration in these organizations and then at their practices in resolving conflict.

For this phase of the study we clearly wanted to select environments that placed different demands on organizations than the plastics industry did. The only available information on which to base this selection was conventional economic data and expert opinion about the natures of various industries. Only after we had chosen the industries were we able to compare them on the environmental measures we used in studying the plastics industry. Therefore we tried to define crudely the major visible factors that gave the plastics industry the diverse and dynamic characteristics we found. The most obvious, as we had mentioned in Chapter II, were the swift technological, scientific, and market changes and the accompanying rapid expansion of the plastics industry. Closely related to these was the fact that the major issue confronting plastics manufacturers was process and product innovation. Finally, since all phases of the environment seemed to bear on this problem of innovation, the basic departments of the company were interdependently involved.

In this second phase of the study, then, we looked for two environments that were different from plastics and also different from each other in at least two of these attributes. One important consideration was to select industries with slower rates of environmental change, which might be more typical of present industrial conditions and less typical of future ones. This comparison, we hoped, would teach us something about the organizational impact of increasing rates of technological and market change. We therefore sought one industry whose rates of growth and change were very slow, and where innovation was not the major competitive factor. For the second environment we wanted an industry where the rates of growth and change seemed to be moderate. While innovation would still be the major issue, we sought an industry in which the capacity to deal with only one part of the environment would be a dominant factor in the organization's success.

After consideration of several alternatives, we chose a segment of the standardized container industry as the most stable environment.* Here the rate of sales increase was only slightly higher than the growth in national population and gross national product over the past five years (Table IV–1). Even more important, no significant new products had been introduced in the past 20 years (Table IV–2). In addition,

* In order to maintain adequate disguise we withhold a more specific designation of the selected segment of the container and food industries. From here on we will refer to the selected segments as the "container" industry and the "food" industry.

Table IV–1

COMPARISON OF POPULATION AND GROSS NATIONAL PRODUCT WITH SALES OF STANDARDIZED CONTAINERS, PACKAGED FOODS, AND PLASTICS INDUSTRY
(1957–1959 = 100)

	1964
Population	106.1
Gross National Product	127.2
Standardized containers	130.0
Packaged foods	138.0
Plastic materials	260.4

SOURCE: Census of Manufacturers; Annual Survey of Manufacturers; Annual McGraw-Hill Survey of Business Plans for New Plants and Equipment, 1965–1968, April 1965; and National Science Foundation, *Basic Research, Applied Research, and Development in Industry*, 1961.

we learned from persons familiar with the industry that the major competitive factors were the operational issues of maintaining customer service through prompt delivery and consistent product quality while minimizing operating costs. The major customers had a large stake in keeping down the

Table IV–2

CURRENT PERCENT OF INDUSTRY SALES COMPRISING PRODUCTS INTRODUCED COMMERCIALLY 5, 10, 20 YEARS AGO
(AS OF 1965)

Industry	*1960*	*1955*	*1945*
Plastics [a]	15%	20%	65%
Packaged foods [b]	5	10	85
Standardized containers [c]	0	0	100

[a] U.S. Tariff Commission Reports; Modern Plastics; Arthur D. Little, Inc.; Standard and Poor's.

[b] Annual Survey of Manufacturers, Standard and Poor's, and Progressive Grocer.

[c] Standard and Poor's.

number of innovations. They wanted to use proven containers that could be processed at high speed on automatic packaging lines, with no delays because of faulty containers. They did not want their existing equipment to become obsolete. For them the container was a part of the product that they wanted to handle with minimum effort.

For our third environment we chose a segment of the packaged food industry. While this industry's rate of growth as a whole was not much higher than that of containers (Table IV–1), we selected a specialized portion of the packaged food industry, where growth had been greater than the industry average. Furthermore, new products in the packaged food industry in general, and in our segment in particular, had been introduced faster than in the container industry (Table IV–2). Both the growth rate of sales and the rate of new product introduction were slower in the food industry than in the plastics industry. Therefore, we ended up with three environments with an apparent range of both growth rates and rates of technological and market change. In the food environment, as in the plastics industry, innovation was a major issue. There is, however, an important difference: In the food industry the market aspect of the environment ("the consumer") was clearly crucial. This fact could place some different demands on the food organizations from those we found in the plastics environment.

These, then, were our initial reasons for selecting these two environments. Let us now turn to a more detailed examination of the characteristics of these two environments and their demands on organizations as compared with the demands of the plastics environment. We will then examine the relationship between these environmental characteristics and the organizational states of differentiation and intergration.

Environmental Demands

To learn more about the demands of these two environments, we will follow the same practice that we did in the plastics industry, using data from both questionnaires and interviews with top executives. We will focus on four aspects of these environments as compared to the plastics environment. Were we correct in assessing the major competitive issue in each industry? How certain is knowledge in the market, techno-economic, and scientific sectors of each environment? How diverse are the characteristics of these parts of each environment? Finally, how tight is the required interdependence of activities in different parts of the environment? From the answers to these questions we hope to develop a clear understanding of the required states of differentiation and integration in each industry, as well as an understanding of the requirements for effective conflict resolution.

As we begin this examination of the environmental demands of these three industries, we should stress again that our measures of these variables are necessarily crude. They are best understood in comparative terms—how does the food industry compare with containers, and containers with plastics, and so on? The practicing manager, as a result of this type of comparison, should be able roughly to fit the organizations with which he has had experience somewhere along the spectrum from the highly dynamic plastics environment to the relatively stable container environment.

The Major Competitive Issues

By interviewing top executives in the food industry we very quickly confirmed our hunch that the major competitive issue in foods, as in plastics, was innovation. Similarly, interviews with container executives confirmed that their main competitive issue was the ability to provide customer service through rapid and timely deliveries and to maintain consistent product quality. Top executives in the food companies described the importance of innovation in this way:

> The big thing in this industry is the almost fanatical desire for new, new, new. This has become the life blood of the business. . . . Prices don't make any difference, and they have no relationship to volume, costs, or anything.
>
> * * * * *
>
> This is a profitable business, which is an intensely competitive market, but not a very price-sensitive one. The top competition takes the form of a very intensive merchandising effort around new product innovations.

In contrast, container top executives discussed the importance of delivery schedules in their industry:

> Our job is to shape the material and wrap it around air, and the only way you can make dollars is to keep moving the product out the door. What happens is that we get the inventories dumped on us, as the customers won't store containers. The paradox is that you have to hold the inventories down, yet you can't make money unless you run the machines constantly. Therefore, sales and manufacturing are constantly at each other's throats. The integration of these conflicting circumstances is the critical management job. . . . The big issue is scheduling, and it is a matter of every day, all day. We either have too much product or too little.
>
> * * * * *
>
> As far as this business is concerned, there is no innovation. If you really want to grow in this business, you have got to have strategically located plants, not giant ones but small ones, well placed throughout the country to give instant service. You have just got to have good delivery service to the customer, optimizing the flow of your material into his plant. Because of the importance of this service, we have become in this business the biggest warehousing industry in the world.

These same container executives also emphasized the importance of product quality as a competitive factor:

> Prices are important in this industry only in the sense that you must meet them. Also, product specifications are standardized, as your product has to be interchangeable with other suppliers'. So we are producing a very undifferentiated product. Obviously, you have to sell something else. This is where you get into the fine point of a quality container in a high-volume business. It can become rather exacting. I think whether you get new business honestly comes down to what kind of container you deliver. The customers, because of the speed at which they run their lines, are very concerned about imperfect containers. They keep detailed records of their losses and whose containers caused them.
>
> * * * * *
>
> Even though the majority of the industry's production is to industry-wide standards, there are still wide variations in product quality. With the customer processing upward of 850 units per minute, any failure of a container impairs their ability to operate. So you must operate with very few defects. You know you are going to have bad containers, but the question is, where do you stack up against your competitor?

As some of these comments suggest, the difference in the major competitive issue between the container environment and the plastics and food environment directly affects both the nature of integration and the degree of differentiation required among organizational units.

Required Differentiation

From our findings in the plastics industry, we expected that one environmental characteristic that would affect the required state of departmental differentiation would be the relative certainty of scientific, market, and techno-economic parts of each environment. The fact that the major issues in the container industry were delivery and quality control, while that in foods and plastics was innovation, suggests that the certainty of the parts of the container environment might be greater than in the other industries. Based on our measures of environmental uncertainty, this was in fact true (Figure IV–1).* All sectors of the container environment were as certain as or more certain than similar sectors of the other two environments. The only aspect of the container environment similar to the others in certainty was the techno-economic portion. This is not surprising, since in all three environments the technology was of a processing type, where

Figure IV–1

RELATIVE UNCERTAINTY OF ENVIRONMENTAL SECTORS [a]

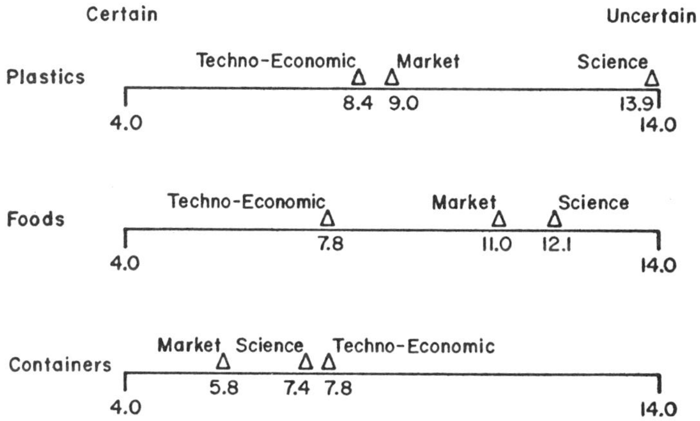

[a] Higher score indicates more uncertainty.

once the process was established, knowledge about it was highly predictable.[1]

The interviews with container executives provide further understanding of the relative environmental certainty in this industry. The remarks of one marketing executive illustrate this point with regard to the container industry market:

* To measure environmental certainty we used the components (clarity of information; certainty of causal relationships, and timespan of definitive feedback) described in Chapter II. Only the total score will be referred to in the present discussion. While the differences between certainty scores are not highly significant, they are consistently in the predicted direction and thus offer empirical confirmation of the data gathered in interviews.

> We and our competitors all use the same machines produced by the same company. The product has got to be virtually the same as your competitor's unless your process is off, as you produce to the same specifications. Nor is there any price competition in this industry. So all we have to know is: How many cases does the customer want, and when?

It is significant that in the container industry market information was more certain than techno-economic knowledge. In the plastics industry, it will be remembered, the situation was exactly reversed, because of the diversity of and constant changes in customer requirements. But while executives in the container industry stressed that they could readily get unambiguous information about how well they were serving their customers, food industry executives emphasized the difficulty of getting clear-cut market feedback. How well a product was performing could only be determined after lengthy test-marketing.

Similar differences existed in the scientific parts of the three environments. In the container industry the necessary scientific knowledge was relatively well understood, and causal relationships were clearly defined. As one container executive put it, most research activity involved solving immediate technical problems where results were quickly apparent:

> If you really examine research in this business, it is very difficult to separate it from quality control. Most of our work is concerned with processing, not new containers. . . . We spend most of our time with manufacturing to keep yields up, and building and designing quality-control equipment.

In contrast, scientific knowledge in the plastics industry was, as we know, highly uncertain, with feedback often coming only after lengthy exploration and experimentation.

While these comments and the empirical data suggest that the parts of the container environment were more certain than those of either food or plastics, the contrast between the latter two industries is not so clear cut. The scientific and techno-economic parts of the environment were apparently less certain in the plastics industry than in foods (Figure IV–1). This seems attributable to the more rapidly changing scientific knowledge in plastics, which also created some technological uncertainties for the production people. Even more interesting was the higher uncertainty of the market in foods as compared with plastics. This difference is consistent with the data gathered in interviews. Plastics executives recognized a number of uncertainties in their market, but they had comparatively few customers from whom they could easily find out the results of a particular action. In the foods industry, on the other hand, there were so *many* customers that the only way to get very firm feedback was to test-market a new product, which took considerable time. Even the causal factors underlying results of these tests were not always clear.

The most important fact that we glean from this comparison of the relative certainty of the various parts of these three environments is that the parts of the plastics and foods environment are more diverse than those in the container industry. The certainty scores are rather similar in all parts of the container environments; information is fairly clear, and causal relationships are well understood throughout this environment. In contrast, in plastics and foods the techno-economic portion of the environment is relatively certain, but the other parts are less so to varying degrees. This lack of diversity of certainty in all phases of the container environment suggests that effective organizations in this industry would not be required to be as differentiated in structure and interpersonal orientation as organizations in the other two industries. The degree of certainty, however, is only one dimension along which the environment can affect the degree of differentiation required of organizations. Let us now examine two other aspects of these three environments and relate them to the state of differentiation that we predicted would be required for effective performance in each industry.

A second factor that influences the required differentiation of units is suggested by the characteristics of the food industry. This is the extent to which one part of the environment overrides the others in importance. In an industry like the food industry, we speculated that the importance of the market could be so great that the functional units of an organization would be required to have less differentiated goal orientation than the units in an industry like plastics, where the elements of the environment are more balanced in importance. With the market so crucial to the organization's success in the food environment, managers in all functional units, not just those in the marketing or sales departments, would be greatly concerned with market goals. As a consequence, the required differentiation among organizational units would be reduced.

The final factor in which we were interested is the required degree of differentiation in time orientation. Our findings in the plastics industry indicated that this would be related directly to the time span of feedback in the different parts of each environment. The data collected on this point clearly suggest that the required differentiation in time orientation was highest in the plastics industry, followed by foods and containers (Table IV–3). The time span of feedback in the container environment was fairly similar in all parts, ranging between just one month and one week. In con-

Table IV–3

TIME SPAN OF FEEDBACK FROM ENVIRONMENT

Industry	*Science*	*Market*	*Techno-economic*
Plastics	1 year	1 month	1 month
Food	6 months	6 months	1 week
Containers	1 month	1 month	1 week

trast, the time span of feedback in plastics ranged from one year for the scientific sector to one month for the techno-economic. Managers in the different units in a container organization, then, would all tend to be oriented toward the same short time horizons, as compared to the quite different time outlooks described in Chapter II for plastics managers. Again the food industry seemed to fall between the other two, with the required differentiation in time orientation less than in plastics but greater than in containers.

We have summarized the state of differentiation required by each environment in Table IV–4.* It seems quite apparent from these data that the container environment required less differentiation of organizational parts than either plastics or

Table IV–4

REQUIRED DIFFERENTIATION [a]

Units	*Required goal differentiation*	*Required time differentiation*	*Required structure and interpersonal differentiation* [b]	*Total*
Sales–Research				
Plastics	2	5	5	12
Food	1	1	1	3
Containers	5	2	2	9
Sales–Production				
Plastics	3	1	1	5
Food	3	4	3	10
Containers	2	1	2	5
Research–Production				
Plastics	5	5	5	15
Foods	4	5	4	13
Containers	1	3	1	5

[a] High score means high required difference.

[b] Since the required difference among units in both these attributes is based on differences in certainty among parts of the environment, these requirements have been treated together.

* An explanation of these scores and their calculation is provided in the Methodological Appendix.

foods, while foods required less than plastics. We therefore predicted that the actual degree of differentiation found in organizations in these three industries would vary accordingly. We were also, of course, predicting that the high-performing organizations in each industry would more clearly meet the environmental requirements for differentiation than their less effective counterparts. Before we test these predictions, we shall turn to the other major environmental demand on organizations, the required degree of integration.

Required Integration

In our investigation of the plastics industry we found that organizations in this environment were required to achieve a high degree of integration. Implicitly, we were assuming that the requirement for integration in this industry might be higher than that in some other industries. The data collected from top executives in the food and container industries indicated, however, that the degree of integration required among different units was virtually the same in all three environments (Table IV–5).

Table IV–5

REQUIRED INTEGRATION [a]

Industry	*Score*
Plastics	5.4
Foods	5.4
Containers	5.5

[a] Higher scores indicate higher required integration.

These scores, however, reflect only the *intensity* of integration required among units, *i.e.,* how dependent one unit's activity is on those of others. They do not reflect some important differences in the *nature* of the integration. In the food industry, where innovation was the major issue in a relatively uncertain environment, the integration had to be brought about in relation to some very complex and uncertain problems. This uncertainty meant that much of the integration (and conflict resolution) must be carried out at the lower levels of the organization, where the required knowledge and information were available. This is similar to the situation in the plastics industry.

In contrast, in the container industry, where uncertainties were fewer and the dominant issues were delivery and quality, the required integration centered on more routine problems and was less frequent and less complicated. This, plus the fact that scheduling decisions affected all plant and sales locations, suggests that the knowledge required for interdepartmental decisions must be centrally processed and could be effectively handled by fewer managers. Since the positional influence for such decisions rested at the top of the organizational hierarchy, it would seem most efficient to collect the required knowledge at this level so that conflicts could be resolved and integration achieved by the upper managers, who had the positional influence and could acquire all the knowledge to do so. There were, however, some other issues, particularly those dealing with product quality, that require detailed knowledge of the production process. These could be more efficiently handled by managers lower in the hierarchy. The characteristics of this environment, however, suggest that in effective organizations integration would usually take place at higher levels.

There was another important difference in required integration for containers as opposed to plastics and foods. Container companies must achieve the tightest integration, around the critical issues of delivery and quality, between sales and production units and also between production and research units. One container executive described the importance of the latter in this way:

With research and production, the cooperation has got to be very close. In fact, it is often hard to tell where one leaves off and the other begins. Those guys in research need to be as interested in the daily figures as I am.

Another container executive described the required integration between sales and production:

We know the industry is going to be in an essentially sold-out condition for the next five years. So you want to service the customers which give you the best profit margins, since you can only make so many containers. This is where the relation between sales and production is critical. If only sales interests schedule the machines, they dance to only the customer's needs. If just the manufacturing interests are reflected, you lose all your flexibility. Both interests have to be represented and evaluated.

In contrast, in the food environment (like the plastics industry) the critical linking relationships were between production and research and between sales and research. We have discussed the reasons for this in the plastics industry; the comments of a food executive explain the importance of these relationships in his industry:

In this business new product ideas come easy. The problem is practicality from both a market and a technical standpoint. So the feasibility of a new idea has to be built in. The key guys are some of the people in research, but typically they don't have the market information or orientation and a lot of inputs must come from [sales], who know the realities of the customer.

Scale-up problems can be tremendous also. What comes out of the commercial plants can have little relationship to what the lab has made. Consequently, a good working relationship between production engineers and the development engineers in the lab is very important.

In summary, while the degree of integration required in all environments was similar, there were important differences in its nature. In plastics and foods it revolved around the complex and uncertain problems of innovation. Thus the research unit must be closely tied in with both production and sales. In the container environment, where integration centered on more certain and programmable issues, production must be closely related to both research and marketing. Because of the differences in issues and in the certainty of the various elements of these environments, integration must usually occur at the upper levels of the container organizations, and at lower levels in the food and plastics organizations.

We must also recognize that, because differentiation and integration are essentially antagonistic, the higher degree of required differentiation, particularly in the plastics environment but also in foods, suggests that the problems of achieving integration will be greater in these industries than in the container industry. The importance of this fact will become more evident after we have examined the actual states of differentiation and integration that we found in the two food organizations and the two container organizations.

Differentiation, Integration, and Performance in the Three Environments

Before exploring how well the organizations in the container and food industries met the requirements of their environments, and how they compared with the plastics organizations described earlier, we shall briefly describe the four organizations investigated in this phase of the study and their relative performance.

Container and Food Organizations and Their Performance

Both the container and the food organizations were major product segments of larger corporations. In each industry the two organizations studied were direct competitors, producing similar, if not identical, products, which were sold to the same markets. The two container organizations had the same basic functional units as plastics firms (research, production, and sales), except that there was only one research unit in each container organization. One of the container organizations (the low performer) also had an integrating unit, which had the prescribed responsibility of integrating sales and production around scheduling and customer service issues. The high-performing organization had no formal integrating unit.

Each of the two food organizations had a production unit and only one research unit, but each had two units assigned the responsibility of coping with market matters. One of these, which we have labeled the marketing department, was involved in formulating and executing new product, pricing, promotional, and advertising plans. The other, which we call the sales unit, was involved in distribution—calling on wholesale and retail stores to insure proper product display and to encourage the stores to maintain adequate stocks. Since the marketing department was the unit primarily involved in product innovation, we shall focus our attention on it. The fact that the managements had made this further division of labor to deal with the market seems to be one result of the dominance and uncertainty of this phase of the environment in the foods industry. In contrast, the plastics organizations elaborated additional units to deal with scientific problems—the most uncertain aspect of their environment. This suggests, perhaps, that organizations will tend to elaborate and subdivide units that cope with the more problematic or uncertain sectors of their environments.

The high-performing food organization had no integrating unit, while the low-performing company had one, which was assigned the responsibility of integrating the activities of marketing and research and production and research around issues related to product innovation.

As the nomenclature above suggests, we have attempted to obtain contrasts between an economically successful organization and a lower-performing one in each of these environments. In the container industry we were able to find the clear contrast we were seeking. The high-performing organization had a higher rate of growth in profits and sales than the low performer. The managers' subjective appraisal of the performance of each organization was in the same direction. The chief executive of the high-performing organization rated his company as closer to his performance ideal than did the chief executive of the low-performing firm.

The clear difference in performance between these two organizations was also evident from interviews with other top managers in both organizations. The executives in the high-performing organization were highly satisfied with their ability to compete in this industry. They felt that all parts of their organization were effectively performing their tasks. In contrast, the top executives interviewed in the low-performing organization were concerned about their present level of performance. They saw a need for improvement in customer service and meeting delivery schedules in order to raise their competitive position.

The contrast in performance between the two food organizations was not so clear. Our two organizations were initially selected for study because of the differences in their general reputations for effectiveness. However, a closer investigation of their relative growth in sales and profits indicated that the low-performing organization was currently growing faster than the high performer. The reason for this was that the low-performing organization had recently had a dramatic success with a new product. In other conventional measures of economic performance (market share, return on investment, number of new products introduced over a five-year period), however, the high performer was doing better than the low performer.

After looking more carefully at these performance data of the food organizations, we recognized that we had chosen as the low performer an organization that had been doing poorly for a number of years, but that was now doing better. While its effectiveness was still not so high as that of the higher performer, it was improving rapidly. This conclusion was borne out by both the subjective evaluations of the chief executives and interviews with other top managers in each organization. The two chief executives' appraisals indicated that the organization selected as the high performer was more effective, but that the difference was not so great as that between the container organizations. The top executives in the high-performing food organization, we learned, were well satisfied with their present level of performance, but were still anxious to improve it (partly because they were concerned about the improving performance of competitors like the low performer). The top executives in the low-performing organization indicated that while they were highly pleased with their recent progress, they were still not satisfied and saw the need for greater effort if they were to match the performance of more effective organizations.

As we discuss the differentiation and integration achieved by the two food organizations, then, we will want to bear in mind that the differences in their performance records were not so dramatic as those between the high- and low-performing container or plastics organizations. In fact, their relative performance more nearly matched that of the high- and medium-performing plastics firms.

Organizational States and Environmental Demands

In examining the actual states of differentiation and integration achieved in these four organizations, we shall make a two-way comparison. We shall compare the high-performing organizations in all three environments to learn whether the actual differences in the states of differentiation and integration matched the environmental requirements.* Are there differences among these organizations that are related to the varying environmental demands? Also, we shall compare the high- and low-performing container and food organizations to see whether our findings from the plastics environment were confirmed in these other two industries. Do the high-performing organizations conform more nearly to the demands of the environment than the low-performing ones?

As we had expected from the variations in required differentiation in these environments, we found that the high-performing plastics organization was most highly differentiated of the three high performers (Table IV–6). Similarly, the high-performing food organization was more highly differentiated than the high-performing container organization.

Table IV–6

AVERAGE DIFFERENTIATION AND INTEGRATION ACROSS THREE ENVIRONMENTS [a]

Industry	*Organization*	*Average differentiation*	*Average integration*
Plastics	High performer	10.7	5.6
	Low performer	9.0	5.1
Foods	High performer	8.0	5.3
	Low performer	6.5	5.0
Containers	High performer	5.7	5.7
	Low performer	5.7	4.8

[a] Comparable pairs of units are used in all six organizations (sales–production, production–research, and sales–research). The average scores reported here for plastics organizations are thus slightly different from those reported in Chapters II and III, since fundamental research and integrative units have been excluded to achieve comparability. Higher differentiation scores mean greater differences. Higher integration scores mean better integration.

* To simplify this part of our discussion we have selected one of the high-performing plastics organizations (A) and one of the low-performing plastics organizations (A) to compare with the food and container organizations. These two organizations in most characteristics represented the extreme organizational conditions found in the plastics industry.

This same pattern is also evident among the low-performing organizations in the three industries. Not only were these differences reflected in these average differentiation scores, but the differentiation between each pair of units in the plastics organizations was greater than that of the comparable pairs in the two food organizations, which were more highly differentiated than their counterparts in the container organization.

As a further check on whether the high-performing organizations actually met the requirements for differentiation, we also found that there was a significant rank-order relationship between the actual and the required degree of differentiation for each pair of units in the three high-performing organizations.* The more the parts of the environment differed in certainty and timespan of feedback, and the less dominant any one part was, the more differentiated were the pairs of units in the high-performing organizations.

This particular relationship between actual and required differentiation did not hold for the three low-performing organizations. This suggests that the low-performing organizations did not meet the requirements imposed by their respective environments so well as their high-performing competitors. This conclusion is supported, at least for the plastics and food industries, by two other pieces of evidence. In these two industries, the high-performing organizations were in total more highly differentiated than their less effective competitors (Table IV–6). Further, there was a significantly closer fit between the attributes of each unit and the demands of its relevant part of the environment in the high-performing food organization than in the low performer (Table IV–7). This corroborates similar findings for the plastics organization (see Chapter II). Thus in both of these industries the high-performing organizations had units whose members' ways of thinking and organizational practices were consistent

* Spearman's rank-order correlations—.61 which is significant at .05.

Table IV–7

COMPARISON OF UNITS' REQUIRED AND ACTUAL ATTRIBUTES

	Number of Positive Comparisons:	
Industry	*High-Performing Firms*	*Low-Performing Firms*
Foods	10 [a]	2 [a]
Containers	8	4

[a] Difference significant at .05 (Fisher's exact test).

with the demands of their particular part of the environment. This did not seem to be so true in the low-performing organizations in these two industries.

In the container industry the two organizations were more nearly similar in their states of differentiation. There is, however, some evidence (Table IV–7) that the high-performing organization's units more closely fitted their part of the environment than did the units in the low-performing organization. Although this difference is not significant, it is in the predicted direction.

Thus the high-performing organizations in the container and food industries, like those in plastics, had achieved states of differentiation that met the demands of their particular environments. While the low-performing organizations (with the exception of the low-performing container organization) had not attained this close a match, the variations between their states of differentiation suggest that they were responding to some extent to their particular environments. But, as we suggested in Chapter II, their failure to realize the required state of differentiation was probably reducing the ability of each unit to deal with its own part of the environment.

In more concrete terms this means that in the high-performing container organization sales, production, and research managers were more similar in their ways of thinking, and their departments tended to follow organizational practices that were not much different. In contrast, as we saw in

Chapter II, the functional managers in the high-performing plastics organizations were quite different in their ways of thinking, and their units' structures varied to fit the differences in their tasks. Thus each high-performing organization was differentiated according to the demands of its environment, and this seemed to contribute to its effectiveness.

The high-performing organizations also seem to have achieved states of integration consistent with environmental requirements. In all three industries the high-performing organizations had better integration than their low-performing competitors (Table IV–6) . Not only were the effective organizations more nearly achieving the required states of differentiation, but they were also generally obtaining more effective working relationships among their units. The one partial qualification to this statement is found in the high-performing food organization.

Since the environmental requirements for closeness of integration were the same in all three industries, we expected that the high-performing organizations would also have to achieve approximately the same states of integration. The high-performing container and plastics organizations were almost identical in this respect. While the more successful food organization was achieving better integration than the low performer, it was not integrating its units so well as the high-performing organizations in the other industries. Moreover, there is evidence from interview comments as well as from the integration scores that this organization faced some problems in achieving integration the other two high performers did not have. In particular, the relations between research and marketing were posing some problems. As one research executive described the situation:

> I don't show these products to marketing. In fact, I never show it to marketing until all our own people have seen it. I couldn't care less about marketing. They more or less like what we tell them to like. . . .

Another research executive said:

> When some ideas come from marketing, I kind of wonder about these things, and I don't trust the ideas. They aren't as well established as we are, and we aren't as tolerant of their mistakes as we are of our own.

Marketing personnel made similar comments about the relationship with research. One marketing manager expressed the general concern that research and marketing did not work closely enough together:

> The basic dilemma we have is the separation of research and marketing. However, we have been steadily hacking away at this as far as research is concerned, and I think we are beginning to make some progress.

Another marketing manager expressed the view that the problems between marketing and research were often at the upper echelons:

> I would say the biggest problem is the feeling that nobody [from outside] is going to direct research. This gets in the way of most spontaneous cooperation between research and marketing. . . . This is not so much a problem at the working level, but at the higher administrative levels.

While the problems of achieving integration were more serious at the upper levels of the managerial hierarchy, there was some evidence of difficulty at several levels.

These and similar comments, together with the integration scores, suggest that while this organization was achieving a more satisfactory state of integration than its low-performing competitor, it was not so well integrated as the high performers in other industries. In spite of this it seems clear that all the high-performing organizations maintained states of integration that more clearly met the demands of their specific environments than those of their low-performing competitors.

CONCLUSIONS

The findings reported in this chapter suggest that the states of differentiation and integration in effective organizations will differ, depending on the demands of the particular environment. In a more diverse and dynamic field, such as the plastics industry, effective organizations have to be highly differentiated and highly integrated. In a more stable and less diverse environment, like the container industry, effective organizations have to be less differentiated, but they must still achieve a high degree of integration.

From these findings we begin to get a partial answer to our basic research question of what organizational characteristics fit different environmental conditions. These differences in states of differentiation and integration are, however, only part of the answer, because these two states are antagonistic. The more differentiated an organization, the more difficult it is to resolve conflicting points of view and achieve effective collaboration. This suggests that organizations in the plastics industry would have more problems in achieving integration than those in the food industry, and that both of these would have a harder time than the container organizations.

The problems of achieving integration in the plastics or food organizations would be further complicated by the relative uncertainty of their environment compared with that of the container industry, and by the fact that, as we mentioned earlier, they have to integrate around complex issues of innovation while the container organizations cope with more certain operational issues. This suggests that the devices and practices used to resolve conflict and achieve integration might differ among the organizations in these three environments. The way these methods and practices varied among these organizations throws further light on our central research question. It is to these differences that we now turn.

CHAPTER V

Additional Perspectives on Resolving Interdepartmental Conflict

In Chapter III we found important differences in the ways in which high- and low-performing plastics organizations handled interdepartmental conflict and decision making. Now we want to see whether similar differences occurred between the organizations in the food and container environments. Do the determinants of effective conflict resolution vary between the high- and low-performing organizations in these two industries? Could these differences account for the ability of the high-performing organization to achieve both the differentiation and the integration required by its particular environment?

In this discussion we must review the point made in Chapter IV, that there were important differences among the three industries that affected the requirements for effective conflict resolution. In the container industry conditions were more certain in all parts of the environment, and the knowledge required to make interdepartmental decisions was available only at the upper levels in all departments. Only the higher executives knew enough about both total market demand and available plant capacity to make the crucial scheduling and quality decisions. In the food industry, on the other hand, the scientific and market parts of the environment were less certain, and the knowledge needed to make decisions about the important issues of product innovation resided in the middle and lower managerial levels of the research and sales departments. Techno-economic knowledge was more certain, so the information required for conflict resolution was available at higher levels in the production department.

A second point that should be reiterated here is that the major issues confronting the organizations in each environment also imposed particular requirements for effective handling of conflict. In the container organizations, where operational issues of customer delivery and product quality were crucial, managers in the sales and production departments were required to have the most influence in resolving interdepartmental conflict. In the food industry, where knowledge about the market and about food sciences was critical, the sales and research units had to be the most influential.

The Process of Conflict Resolution in the Container Organizations

Integrative Devices

Only the low-performing container organization had a formal integrating unit. This department, which reported to the general manager, had as its assigned function the integration of sales requirements and production capacity at the organization's several plants. In addition, the responsibility for scheduling decisions involving any one plant had been assigned to the individual plant managers and their counterparts in the sales department (the regional sales managers). As the reader may well suspect, these arrangements led to considerable confusion about where conflicts were to be resolved. We shall shortly explore this problem in more detail.

In the high-performing container organization all integration was carried out through the managerial hierarchy. No special departments, teams, or roles, other than a scheduling clerk, had been established to facilitate the process of conflict resolution. In this sense this company approached the prototype of the *classical* organizational model, where all decisions should be made through the formal hierarchy. As we examine the determinants of effective conflict resolution in these two organizations, we shall see how this particular model fitted the requirements of this relatively stable and homogeneous environment.

The Determinants of Effective Conflict Resolution

The first factor that seemed to determine each of these organization's capacity to handle interdepartmental conflict was the relative influence of the various departments. In both companies the sales and production departments apparently had the highest influence, and the former was significantly more influential than the latter (Table V–1).* Generally this was consistent with the demands of the environment since both departments had to be involved in the crucial scheduling decisions and since it was the sales department that had close contact with the customer and had to initiate scheduling demands on production. We predicted, however, that if this inequity in influence became too large, the production people might feel that their point of view was not being given sufficient attention. If this happened, conflicts would be more difficult to resolve constructively.[1]

Table V–1

DEPARTMENTAL INFLUENCE IN THE CONTAINER ORGANIZATIONS[a]

Department	*High performer*	*Low performer*
Sales	3.4	4.0
Production	2.5	2.8
Research	2.0	1.9

[a] Higher number means higher influence.

* The difference between the sales and production units was significant at .05 in the high-performing organization and .01 in the low-performing organization (Neuman-Keul's Test for Ordered Differences).

In the high-performing organization, where this inequity was smaller than in the low performer, the production managers indicated that they felt their problems of plant scheduling were given sufficient attention. This may have been partially because, as we shall see, most of these decisions were made by upper management, who were seen as knowledgeable in both plant and markets matters.

In the low-performing organization, however, the imbalance of influence between production and sales was larger, and production managers indicated a dissatisfaction with their ability to influence decisions.* Two quotations from production managers will illustrate this point:

> Scheduling is now a nip-and-tuck game with sales, and as far as I am concerned sales runs the plant. For example, if they want their items next week, we just have to do it; that's the way the game is played around here.
>
> * * * * *
>
> I believe more time should be spent by salesmen in the plant and in the office. I think they ought to make regular visits to see our problems. They don't appreciate our problems. Of course, the salesman's job is to take care of the customer, so they do everything they can to see that the product is made and put in stock, so the customer can get it whenever he wants it. However, they don't seem to realize the problems this creates in the plants. More visits and more contact would improve this greatly.

As these comments suggest, production managers in the low-performing organization felt that they had too little influence over decisions. This imbalance in influence seemed to reduce the effectiveness of integration between the two units

* The differences in influence between the two sales and between the two production units were not significant, but as mentioned above, the differences between the sales and production units in each organization were statistically reliable. This suggests that any real difference between the two organizations was in the inequity in influence between sales and production. In any case, the differences were in the predicted direction.

which had the knowledge needed to resolve the important interdepartmental conflicts around issues of customer service.

The second determinant of effective conflict resolution that we explored was the total influence of all members of the organization. Based on our findings in the plastics industry, we would expect that the total influence would be greater in the high-performing organization than in the low-performing one, but this turned out not to be the case (Table V–2). The managers in the low-performing organization actually felt that they had significantly more influence on decisions than did those in the high-performing company.

Table V–2

TOTAL INFLUENCE IN THE CONTAINER ORGANIZATIONS

Organization	*Total Influence*
High performer	2.6 [a]
Low performer	3.3 [a]

[a] Significantly different at .05. High score indicates high influence.

According to our earlier findings, this high total influence would suggest that the managers in the low-performing container organization should be able to work more effectively together in resolving interdepartmental conflicts. However, in extrapolating this conclusion from the plastics organizations to the container industry, we are generalizing about the one best organizational form without recognizing the differences in environmental demands in the two industries. In the plastics organizations the middle-level and lower-level managers had the knowledge required to make effective decisions, but in the container industry much of the required knowledge rested only at the upper levels of the three basic departments. While the managers in the low-performing container organization may have felt that they had important influence over decisions, they did not have the knowledge necessary to participate effectively in resolving conflicts and reaching these decisions.

The reader will recall that in discussing this matter of total influence in the plastics companies we offered two possible reasons why high total influence could contribute to effective conflict resolution and performance: by improving managerial motivation, or by improving the quality of decisions as the relevant knowledge was brought to bear. Now the evidence from the container companies would indicate that the latter explanation is the more valid. Having the knowledgeable people exert the influence seems to contribute to the quality of conflict resolution, even though it depresses the total influence. This is not to say that it might not also have a lesser side effect in depressing motivation. More will be said about this below.

A clearer understanding of the influence pattern in these two organizations can be gathered from our findings about another determinant of effective conflict resolution—the management level at which influence to resolve conflict was concentrated. In the high-performing container organization the influence seemed to be more concentrated near the top of the hierarchy (Figure V–1). Although the difference in the slope of the lines was not statistically significant, in the low-performing organization the influence line tended to be less steep and influence seemed to be somewhat more evenly distributed. This finding, together with the data on total influence, suggests that while the managers in the high-performing organization generally felt that they had less influence to resolve conflicts, those executives who were seen as having high influence were the top managers, who had the requisite knowledge to make decisions. We are not suggesting that the lower and middle managers in the high-performing container organization were not involved in any decisions. Our interviews indicated that they were particularly concerned with resolving technical and quality problems. So the data

Figure V–1

DISTRIBUTION OF INFLUENCE IN THE CONTAINER ORGANIZATIONS [a]

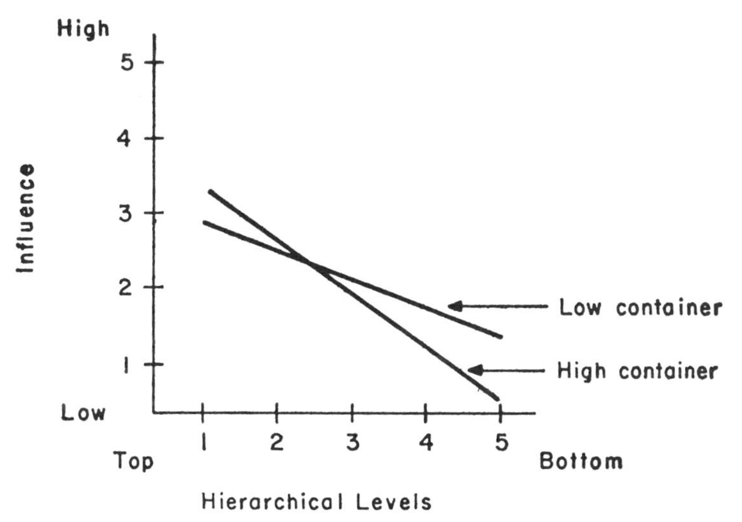

[a] Lines fitted by least square methods. The probability of obtaining the differences in the slopes of these lines by chance is $< .25$.

suggest that they did have some influence in resolving conflicts, though less than their counterparts in the low-performing organization.

These differences between the two organizations are even more dramatically illustrated by the tone of comments from managers about the level at which conflicts were resolved. In the low-performing organization there had been a systematic attempt, as we have mentioned, to get lower and middle managers involved in making decisions about scheduling and other customer-service issues. While this apparently gave them more feeling of having influence over these decisions than managers had in the high-performing organization, it also left them with a feeling of frustration because they were not able to achieve any clear resolution. As one production manager said:

Presently there are thousands of guys involved in the scheduling process around here. It just doesn't work with so many cooks in the stew. . . . What happens now is that when an order comes in it is my decision to take care of it and to schedule it into the first available production space and then give some other orders a later production date. This date usually goes back to the sales office, and they aren't happy, and they will start for the regional manager or the [integrating department] in headquarters to try to get this date improved. Consequently, the [integrating] guys are always dabbling with us about the schedule. This really gets frantic sometimes. On the other hand, there has to be a bit of this in the situation we are in. Hell, I don't know all the facts out here [in an outlying plant]. It is just an order, and we don't know what's behind it or how important it is to the company. This, I think, also accounts for a lot of meddling in the scheduling.

This manager was pointing to the central problem in his organization. The middle-level managers, who were being asked to get involved in resolving scheduling conflicts, did not know enough about plant capacities and customer requirements to make these interdepartmental decisions. As a consequence, the decisions of the regional and plant managers were frequently reversed by the integrating department or, more often, by higher headquarters management. The comment of one salesman describes the dilemma of the regional manager:

My boss, the regional manager, is doing an impossible task. And in fact, what he has been turned into is a production scheduler. He is so swamped with these petty problems that he hasn't made a call with us on customers in weeks. What really happens is that we get one direction from the regional manager and another from the home office, and when we ask the regional manager about it, all he is able to say is that they double-crossed him. . . .

Although this salesman did not know who finally made these decisions, there was little doubt among the managers at the organization's headquarters that conflicts were eventually resolved by the top departmental managers and the general manager. Even though the middle executives began by trying to resolve the conflicts and make decisions, they found that they were not effective because of a lack of information about the total situation. Thus conflicts were sent up the organizational ladder, past the integrating department, to the upper echelon, where adequate knowledge to make decisions was available. The managers at the middle levels were frustrated by their inability to handle these conflicts as they had been led to believe they should, and the problems intensified and festered as they were passed upward.

In contrast to this situation, interviews with managers indicated, as the influence data (Figure V–1) suggested, that in the high-performing organization the influence to resolve conflict was more concentrated at the upper levels. From talking to the managers at all levels in the organization, we could clearly see who made crucial interdepartmental scheduling decisions. It was the chief executive officer, with the help of other principal officers. These men had the required knowledge about the total situation, and they maintained almost complete control over all scheduling decisions. A production executive described the chief executive's involvement in this way:

> In actuality, of course, in this business we are dealing with a balance so delicate that you must be in a high position to appreciate all the factors involved. . . . [The chief executive officer] does all the scheduling in his head, and then tells us what to do. He has scheduling meetings weekly, and then others sit in as a data source. Scheduling is definitely regarded in the sales area. It is not a production thing, and this is a primary consideration. Almost no decisions in this area are production decisions. They don't give production considera-

> tion unless you can't produce something they want to do. This is not to say, though, that sales is unconscious of production problems. Production doesn't sit in on the meetings, though. However, most people who are there are aware of the problems of producing containers and of the plant capacities, so the production interests are not absent, even if the people are not there.

We should also note that this manager and his production department colleagues felt that their interests were well represented in the resolution of conflict by the top executives. This is related to the point made earlier about the relative influence of the sales and production departments in this organization.

One of the top sales executives, who was directly involved in the scheduling decisions, described the process by which they were reached:

> The assignment of the machine schedules is worked out by myself, the sales vice president and [the chief executive], who will sit down to figure this out. We do this at least once a week. What we work at is to recap every item we know of at the present time by product and customer. We sit down to do this once on Mondays, and then throughout the week, as something arises, we will make changes. All day long this is going on, and [the chief executive] is in and out of here all day long. We keep him advised. He is usually talking to me at least two or three times a day.

A salesman's comments provide another perspective on the chief executive's role in the resolution of conflict:

> On new customers the sales vice president tells me whether we can do the business. However, the real person is the chief executive, because he wants to be updated immediately on everything. He wants to be kept up to date on everything, and he does get himself involved in everything. He runs this place completely. And he gets fancy if something happens that he wasn't aware of and wasn't involved in. . . .

> I thought at first when I came here this was meddling. However, now, as I think about it, at least you have a guy who is going to give you decisions. If I want answers quickly, I'll go directly to him, keeping the other people tied in as I go. . . .

This salesman is also reflecting the common attitude among this organization's personnel about the chief executive's extreme control of decisions. While a few managers indicated that they were unhappy because of their own lack of involvement, the vast majority saw the situation as necessary and conducive to their effectively performing their own jobs. When conflicts arose, they knew that they would get them resolved by the top managers, who had both the knowledge of the environment and the influence required.

All these data, then, suggest that one important determinant of the high-performing organization's ability to resolve conflict effectively was the centralization of influence. In the low-performing organization total influence was greater, and more managers felt that they were involved in trying to resolve conflicts. Since they lacked the knowledge to make decisions, however, their involvement in this process led primarily to frustration and increased conflict.

One other related factor that affected the conflict-resolving capacity of the high-performing organization was the *basis* of the influence of the chief executive and his top aides. While one might at first glance suspect that their high influence was based solely on their positions and the fact that they held the purse strings, this was not entirely the case. In interview after interview we heard unsolicited comments describing the chief executive as "a brilliant man"; "a man who knows more about this industry than anyone else"; etc. Similar comments were made about the other principal officers. These remarks, along with some cited earlier, which recognized top management's unique position to have overall knowledge about industry conditions, suggest that the influence of top management was not derived from position alone, but also from knowledge and competence in dealing with industry problems. Although there is no evidence that the top executives in the low-performing organization did not seem competent to their subordinates, it is clear that they did not command the broad respect based on ability that the top management in the high-performing organization did. The fact that the chief executive and his management colleagues in the high-performing organization were seen as being so competent seemed to result in their decisions' being accepted as sound by all parties to conflicts.

This matter of influence based on knowledge and competence was also connected to the relative ineffectiveness of the integrating unit in the low-performing organization. While managers in this unit seemed to have the balanced orientations required for them to be effective integrators, they, like the other middle-level managers, often lacked the necessary knowledge. Beyond this, they also lacked influence. As a manager in the integrating department put it:

> The traditional method of resolving conflict in this company is by taking it upstairs. This often cuts the legs off our own work. It spoils the effort we are making to create an impartial image as a referee. We are not the final authority. Either party can take the matter up further. We don't have the authority to solve these problems at the lower levels.

If this integrating unit had been privy to the information required to make these decisions, and if it had, in the eyes of the functional managers, strong knowledge-based influence, it might have played a useful role. But these are big "if's." In fact, the unit was able to do little more than add further confusion to an already hectic situation.

A final factor related to our present discussion is the mode of behavior used to resolve conflict. In the high-performing organization there was significantly more confrontation than in the low-performing competitor (Table V–3). Judging from

Table V–3

MODES OF CONFLICT RESOLUTION IN THE CONTAINER ORGANIZATIONS

Organization	*Forcing*	*Smoothing*	*Confrontation*
High performer	8.6 [a]	9.1 [a]	12.9 [a]
Low performer	9.1 [a]	8.6 [a]	12.2 [a]

[a] Difference significant at .01. Higher score indicates more typical behavior.

the pattern of hierarchical influence discussed above, much of the confrontation in the high-performing organization was apparently at the upper organizational levels. As we have suggested, however, many technical and quality problems were resolved at lower levels, again through the use of confrontation. These managers were apparently solving their conflicts and reaching decisions by openly examining different points of views. This difference between the high- and low-performing container organizations is consistent with our earlier findings in the plastics industry about the importance of confrontation of conflict for effective decision making.

In the high-performing container company (unlike the high-performing plastics organizations), we found less forcing and more smoothing than in the less successful competitor. However, the relatively large amount of smoothing in this high-performing organization is consistent with the point made earlier—that top executives did not appear to subordinates as capriciously using the power of their position to force compliance, but rather as competent persons using their knowledge to reach decisions. In fact, they seemed to be using smoothing as their back-up mode of handling conflict, perhaps as a way of maintaining good relations when they reached decisions that were particularly unpopular with any one functional unit. Contrary to what we saw in the plastics organizations, smoothing in this case may have been a useful secondary mode. The low-performing organization relied more on forcing. This, too, is consistent with our earlier discussion, which suggested that decisions were passed to higher levels of management, because these executives could reverse a decision that one lower functional manager considered unattractive, even though others saw it as desirable.

From the foregoing comparison we can understand how the high-performing organization was able to achieve better integration (although the same state of differentiation) than its less effective competitor. The more successful company had a better balance of influence between its sales and production units than the low performer. The pattern of hierarchical influence in the high-performing organization was also more consistent with the environmental demands than that in the low performer. Executives at the top of the high-performing organization had both the influence and the knowledge necessary to resolve conflicts, which was not the case in the low performer. We also found that the influence of the top managers in the high-performing organization was at least partially derived from their expertise and knowledge. While to some extent this may have been true in the case of the upper-level managers, in the low-performing organization it clearly was not true for the members of the integrating unit. As a result of their lack of influence and incomplete knowledge, these integrators were not effective in helping to resolve interdepartmental conflict. This suggests that in a stable environment when there is little differentiation among departments, special integrating units are not only redundant, but may actually add confusion. Finally, we found that in the high-performing organization managers were relying more on confrontation and problem-solving behavior as a way of handling conflict than were the managers in the other organization.

As we suggested at the outset of this discussion, the high-performing organization fit the *classical* organizational model, with the influence to make integrating decisions resting heavily on the top managers. Now, however, we can see why this pattern of influence seemed to fit the demands of this fairly stable and homogeneous environment. The attempts to "decentralize" influence in the low performer did not serve these particular environmental conditions so well.

We should, at this point, add a few words about some of the risks and costs associated with adopting this classical organizational form. One important problem, about which the managers in the high-performing organization were concerned, is the matter of managerial succession. What happens when the chief executive retires or is away? Even given the fact that there were other capable managers in this organization, removing a manager with this omnipotence and omniscience would require a period of radical adjustment.

Second, as we suggested earlier, some managers in this organization were concerned about their inability to influence decisions. While this was not a serious problem, it raises the important issue of the personality characteristics of managers who are attracted to this type of organization and can find satisfaction in it. We did not attempt systematically to study this issue, but it is worth noting that the reason that the low-performing organization had attempted to spread influence throughout the organization was so that its managers, particularly the younger ones, would be more involved in decisions. The top executives hoped that this would develop their managerial capacity and would make the company a more attractive place to work. Since this effort to decentralize decision making did not fit the organization's environmental requirements, we might at least speculate whether the real problem in the low performer was the attempt to attract managers whose motivational and other personality characteristics did not fit the demands of the jobs available. While this study cannot provide answers to this sort of issue, this brief discussion does remind us of the necessity to consider not only the fit between the organization's structure and its environment, but also the match between the structure and the predispositions of the members.

Conflict Resolution in the Food Organizations

The two food organizations also differed in the extent to which they met the determinants of effective resolution of interdepartmental conflict. In beginning a discussion of these differences, we should re-emphasize the point made in the last chapter, that the low-performing food organization was rapidly improving its performance. As we compare it here with the high-performing organization, we shall see some evidence that progress was also being made in developing more effective practices for handling interdepartmental conflict.

Integrating Devices

One step that had recently been taken to improve the resolution of interdepartmental conflict in the low-performing organization was the establishment of an integrating department. The members of this unit were charged with integrating the product-innovation efforts of the production, sales, marketing, and research departments. In the high-performing organization no integrating department had been established. Instead, individual integrators in the research and marketing departments had formal responsibility for reaching joint decisions. In addition, this organization placed a great deal of reliance on direct contact among the various unit managers. As we shall see, these integrating devices in the high-performing organization were more effective than those in the low performer, but managers in the less successful organization's integrating unit were learning to perform their function and seemed to be improving their abilities to handle disagreements.

Determinants of Effective Conflict Resolution

In this discussion we want first to examine the factors that were directly related to the behavior of only the integrators in these two organizations. In both organizations these men had relatively balanced orientations and ways of thinking. They were concerned not only with the important market factors in decisions, but also, to an almost equal extent, with the scientific and techno-economic factors. Their time orientations were also balanced between the relatively short-term concerns of production and sales and the longer-term orientation of research and marketing personnel. In both organizations these integrators also felt that they were being rewarded most importantly for the performance of the product group with which they were associated. Thus in these two determinants of effective conflict resolution there were no meaningful differences between the managers primarily involved in integration.

Managers in the high-performing organization saw the individual integrators as having more influence on interdepartmental decisions than their departmental colleagues in research and marketing. In the low-performing organization members of the integrating department were seen as having influence equal to that of the members of the research and marketing units. The basis of this influence did, however, differ somewhat in these two organizations, and this factor seemed to affect the relative ability of the integrators to resolve conflicts. In the high-performing organization the influence of these integrators was seen as stemming directly from their ability and knowledge. Comments such as the following were typical:

> When I come up with a product, I take it over to [an integrator] and say, "Look here, taste it," and I pray that he is going to like it. I don't know what good [product] tastes like, and I have come to the conclusion that nobody else in the

company knows what it tastes like. However, fortunately we have a guy like [integrator] who is a 30-year expert, and who can judge what this thing should taste like. If he says it is mushy, it is mushy, and I go back and work some more, but hell, I couldn't identify what mushy was from anything else.

The basis of influence of the integrating department in the low-performing organization was not so clear. Some members of the organization seemed to feel that the influence of the integrators was derived from their close proximity to the chief executive. As one manager put it:

I think it is important to recognize that the [integrating] department here is an extension of the president, and it really helps him to do a lot of things he wants to do. . . .

Other managers seemed to feel that the integrating group's influence stemmed from the competence of its members. As a member of the marketing department said:

That [integrating] group, their influence goes far beyond the description in the organization chart. [The unit's manager] by nature is a hell of a good marketing man and very experienced in the food business. Consequently, the influence of him and his group extends way beyond the normal boundary.

A comment from a production manager indicates the view that was perhaps most typical. He indicated that he and his colleagues were gaining confidence in the ability of the integrating group:

All of us in production are steamed up by the reorganization of this [integrating unit] and feel we needed this reorganization. We just completed the best year in our history, and I think much of it stems from that fact.

The conclusion we drew from these data was that, while at its inception the integrating group was seen as deriving its influence from its proximity to the chief executive, gradually, as it achieved some success in tying together the efforts of other units, its members were beginning to develop a reputation for their expertise. To the extent that the other managers still felt that their influence was derived from their positions, their effectiveness at resolving interdepartmental conflict was somewhat limited, but this situation was improving.

Turning to the determinants affecting the ability of all the members of the organization to handle conflict effectively, we shall first examine the relative influence of the basic departments in each organization. In both organizations we found that the marketing and research groups had significantly higher influence than production (Table V–4). This was consistent with the dominant issue of innovation, which required that the inputs from the market and scientific sectors be combined to develop successful new products. In the high-performing organization, however, members of both of these units had significantly higher influence than their opposite numbers in the competing organization.

Table V–4

FUNCTIONAL DEPARTMENTAL INFLUENCE IN THE FOOD ORGANIZATIONS [a]

Department	*High performer*	*Low performer*
Production	2.3	2.3
Marketing	4.4 [b]	3.8 [b]
Research	4.5 [b]	3.7 [b]

[a] High score indicates high influence.
[b] Influence in the marketing and research units in the high performer was significantly greater at .01 than in the corresponding unit in the low performer. The differences between marketing and research and production in each organization were significant at .01.

Thus it is not surprising that the total influence in the high-performing organization was significantly higher than in the low-performing organization (Table V–5). To interpret this difference in total influence properly, we must first examine another factor that affected their relative capacity for dealing with interdepartmental disputes.

Table V–5

TOTAL INFLUENCE IN THE FOOD ORGANIZATIONS

High performer	3.8 [a]
Low performer	3.1 [a]

[a] Higher score indicates higher influence. Difference significant at .01.

The two organizations differed in respect to the level at which influence was concentrated (Figure V–2). As we have pointed out, given the uncertainty of knowledge, particularly in the market and scientific sectors of this environment, it seemed necessary for influence to be spread throughout all managerial levels if the managers who were attempting to resolve interdepartmental conflict were to have the knowledge required to do so. In the low-performing organization influence was not evenly spread, but was concentrated in the higher levels. This was especially true at the top of the research hierarchy. Research scientists expressed considerable concern about this fact, indicating that it reduced their ability to become involved in settling interdepartmental problems.

One research scientist expressed this widespread concern:

> In many ways I think this is very frustrating. The scientist calls all the development shots until he comes up with an interesting project. Then you lose touch. All of a sudden the thing is taken out of your hands, and you really don't know what is going on in terms of the meaningful decisions that affect the thing you are working on. A lot of decisions are made which affect your work. For instance, on the concept of advertising. However, you only get this information by faith.

Figure V–2

Distribution of Influence in the Food Organizations [a]

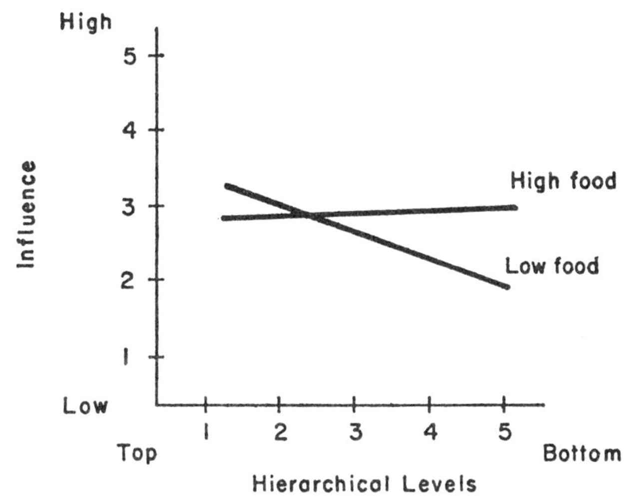

[a] Lines fitted by least square methods. The difference in slope between the two lines is significant at .001.

> You are not tied in. The information is going much higher up, even above the research group manager. I think we should certainly be included in the communication. Now many times the information doesn't filter down to us – information which vitally affects our work.

Not only did this influence pattern create frustration for the working scientists, but it also meant that the research managers involved in decisions did not have available the detailed knowledge they needed. This same difficulty was evident to a lesser extent in the marketing function. Here too influence was concentrated at a level higher than the managers who had the detailed knowledge of markets required for decisions. As we have seen in looking at the organizations in other environments, this lack of congruence between knowledge and influence increased the difficulties involved in resolving interdepartmental conflicts. It suggests again that the differences in total influence mentioned above are another manifestation of the lack of congruence between knowledge and influence. We might point out that the top managers in the low-performing organization were aware of this situation and were working to find methods of spreading the decision making throughout the organization.

Once more the final determinant of effective conflict resolution that we shall examine is the typical mode of behavior used to resolve conflict. Confrontation was more typical in the low-performing organization than in the more successful one (Table V–6). In the other modes of handling conflict the two organizations were similar. Although confrontation was more typical in the low performer, the high-performing organization was doing as much confronting and problem solving as the most effective plastics and container organizations. Since both food organizations compared favorably with all the other organizations in the study, this suggests that the reliance on confrontation should have facilitated the resolution of conflict in both organizations.

Table V–6

MODES OF CONFLICT RESOLUTION IN THE FOOD ORGANIZATIONS

	Forcing	*Smoothing*	*Confrontation*
High performer	8.6	9.6	13.1 [a]
Low performer	8.8	9.6	13.6 [a]

[a] High number means more typical behavior. Difference significant at .01 orthogonal comparison.

While both organizations met this determinant, the differences in the other factors mentioned affected their relative capacity to resolve conflicts and reach interdepartmental decisions. The influence of integrators in the high-performing organization was more clearly based on their competence and expertise than that of the integrators in the low performer. Members of the two functional departments most centrally involved in product innovation (marketing and research) had more influence in the high-performing organization than in the low performer. Finally, the hierarchical pattern of influence was less steep in the high-performing organization than in the low performer. This meant that in the more effective organization managers who had the influence to make decisions and resolve conflicting points of view also had the detailed knowledge of market, scientific, and technical factors required. This was not so clearly the case in the low-performing organization.

Again we have found that the differences in the extent to which two organizations met the determinants for effective conflict resolution help us to understand why the high-performing organization was achieving states of differentiation and integration that more nearly met the demands of its environment than those of its rival. The high-performing food organization, because its managers were dealing with conflict in a way that was consistent with the demands of the environment, was able to achieve both more effective integration and a higher degree of differentiation than its less successful competitor.

Conclusions

In the container and food industries, as in plastics, we have found that the high-performing organizations had methods and practices of handling disagreements that were consistent with the requirements of the particular environment and that seemed to result in effective resolution of conflict. With more effective methods for dealing with conflict, the high-performing organizations tended to achieve the required states of differentiation and integration to a greater degree than their less effective competitors, and the evidence indicated that this contributed to higher performance. This finding suggests an answer to the major question of this study—What kinds of organizations are most effective under different environmental conditions? To be effective, an organization must approach its environmentally required states of differentiation and integration. To achieve these states, it has to meet most of the determinants of effective conflict resolution, so that the managers in the organization can effectively settle the differences that result from their specialized points of view.

This, of course, is a highly abstract and oversimplified answer to our question. In the next chapter we shall elaborate this answer so that the reader will see more precisely how effective organizations in these three environments are similar yet different from one another in relation to the demands of their particular environments.

CHAPTER VI

High-Performing Organizations in Three Environments

IN THIS CHAPTER we shall summarize and amplify the answers we have found to the major question of this study—What types of organizations are most effective under different environmental conditions? By comparing three high-performing organizations we can arrive at a more concise understanding of how their internal differences were related to their ability to deal effectively with different sets of environmental conditions.* This comparison also provides a more complete picture of each organization, to allow the reader to move beyond the numerical measures and gain a fuller appreciation of the distinct characters of these three effective organizations. While our focus will be on the high performers, we shall draw occasionally on our findings about the other organizations for help in clarifying our conclusions.

It may seem, in this summary, that we are describing "ideal types" of organizations, which can cope effectively with different environmental conditions. This inference is not valid for two reasons. First, we believe that the major contribution of this study is not the identification of any "type" of organization that seems to be effective under a particular set of conditions. Rather, it is the increased understanding of a

* We shall continue to use organization A as our example of a high-performing plastics organization, but the reader should be reminded that the other high-performing plastics organization was in all important features quite similar to organization A.

complex set of interrelationships among internal organizational states and processes and external environmental demands. It is these relationships that we shall explicate further in this chapter. Second, although all three high-performing organizations were effective in dealing with their particular environments, it would be naive to assume that they were ideal. Each one had problems. One characteristic that the top managers in these organizations seemed to have in common was a constant search for ways to improve their organization's functioning.

Organizational States and Environmental Demands

In each industry, as we have seen, the high-performing organization came nearer meeting the demands of its environment than its less effective competitors. The most successful organizations tended to maintain states of differentiation and integration consistent with the diversity of the parts of the environment and the required interdependence of these parts. As we indicated in Chapter IV, the differences in the demands of these three environments meant that the high-performing plastics organization was more highly differentiated than the high-performing food organization, which, in turn, was more differentiated than the high-performing container organization. Simultaneously, all three high-performing organizations were achieving approximately the same degree of integration.

To illustrate the varying states of differentiation among these three organizations, we can use hypothetical encounters among managers in both the plastics and the container high-performing organizations. In the plastics organization we might find a sales manager discussing a potential new product with a fundamental research scientist and an integrator. In this discussion the sales manager is concerned with the needs of the customer. What performance characteristics must a new product have to perform in the customer's machinery?

How much can the customer afford to pay? How long can the material be stored without deteriorating? Further, our sales manager, while talking about these matters, may be thinking about more pressing current problems. Should he lower the price on an existing product? Did the material shipped to another customer meet his specifications? Is he going to meet this quarter's sales targets?

In contrast, our fundamental scientist is concerned about a different order of problems. Will this new project provide a scientific challenge? To get the desired result, could he change the molecular structure of a known material without affecting its stability? What difficulties will he encounter in solving these problems? Will this be a more interesting project to work on than another he heard about last week? Will he receive some professional recognition if he is successful in solving the problem? Thus our sales manager and our fundamental scientist not only have quite different goal orientations, but they are thinking about different time dimensions —the sales manager about what's going on today and in the next few months; the scientist, how he will spend the next few years.

But these are not the only ways in which these two specialists are different. The sales manager may be outgoing and concerned with maintaining a warm, friendly relationship with the scientist. He may be put off because the scientist seems withdrawn and disinclined to talk about anything other than the problems in which he is interested. He may also be annoyed that the scientist seems to have such freedom in choosing what he will work on. Furthermore, the scientist is probably often late for appointments, which, from the salesman's point of view, is no way to run a business. Our scientist, for his part, may feel uncomfortable because the salesman seems to be pressing for immediate answers to technical questions that will take a long time to investigate. All these discomforts are concrete manifestations of the relatively wide differences between these two men in respect to their working and thinking styles and the departmental structures to which each is accustomed.

Between these different points of view stands our integrator. If he is effective, he will understand and to some extent share the viewpoints of both specialists and will be working to help them communicate with each other. We do not want to dwell on his role at this point, but the mere fact that he is present is a result of the great differences among specialists in his organization.

In the high-performing container organization we might find a research scientist meeting with a plant manager to determine how to solve a quality problem. The plant manager talks about getting the problem solved as quickly as possible, in order to reduce the spoilage rate. He is probably thinking about how this problem will affect his ability to meet the current production schedule and to operate within cost constraints. The researcher is also seeking an immediate answer to the problem. He is concerned not with its theoretical niceties, but with how he can find an immediate applied solution. What adjustments in materials or machine procedures can he suggest to get the desired effect? In fact, these specialists may share a concern with finding the most feasible solution. They also operate in a similar, short-term time dimension. The differences in their interpersonal style are also not too large. Both are primarily concerned with getting the job done, and neither finds the other's style of behavior strange. They are also accustomed to quite similar organizational practices. Both see that they are rewarded for quite specific short-run accomplishments, and both might be feeling similar pressures from their superiors to get the job done. In essence, these two specialists, while somewhat different in their thinking and behavior patterns, would not find it uncomfortable or difficult to work together in seeking a joint solution to a problem. Thus they would need no integrator.

These two hypothetical examples show clearly that the differentiation in the plastics organization is much greater than in the equally effective container concern. The high-performing food organization fell between the extremes of differentiation represented by the other two organizations. These examples illustrate another important point stressed earlier—that the states of differentiation and integration within any organization are antagonistic. Other things (such as the determinants of conflict resolution) being equal, the more highly differentiated the units of an organization are, the more difficult it will be to achieve integration among them. The implications of this finding for our comparison of these three high-performing organizations should be clear. Achieving integration becomes more problematic as we move from the relatively undifferentiated container organization, past the moderately differentiated food organization, to the highly differentiated plastics organization. The organizational problems of achieving the required states of both differentiation and integration are more difficult for a firm in the plastics industry than for one in the container industry. The next issue on which we shall compare these three organizations, then, is the devices they use to resolve conflict and achieve effective integration in the face of these varying degrees of differentiation.

Integrative Devices

Each of these high-performing organizations used a different combination of devices for achieving integration. As the reader will recall, the plastics organization had established a special department, one of whose primary activities was the integration of effort among the basic functional units (Table VI-1). In addition, this organization had an elaborate set of permanent integrating teams, each made up of members from the various functional units and the integrating

Table VI–1

COMPARISON OF INTEGRATIVE DEVICES IN THREE HIGH–PERFORMING ORGANIZATIONS

	Plastics	*Food*	*Container*
Degree of differentiation [a]	10.7	8.0	5.7
Major integrative devices	(1) Integrative department	(1) Individual integrators	(1) Direct managerial contact
	(2) Permanent cross-functional teams at three levels of management	(2) Temporary cross-functional teams	(2) Managerial hierarchy
	(3) Direct managerial contact	(3) Direct managerial contact	(3) Paper system
	(4) Managerial hierarchy	(4) Managerial hierarchy	
	(5) Paper system	(5) Paper system	

[a] High score means greater actual differentiation.

department. The purpose of these teams was to provide a formal setting in which interdepartmental conflicts could be resolved and decisions reached. Finally, this organization also placed a great deal of reliance on direct contact among managers at all levels, whether or not they were on a formal team, as a further means of reaching joint decisions. As Table VI–1 suggests, this organization, the most highly differentiated of the three high performers, had the most elaborate set of formal mechanisms for achieving integration and in addition also relied heavily on direct contact between managers.

The food organization had somewhat less complex formal integrative devices. Managers within the various functional departments were assigned integrating roles. Occasionally, when the need for collaboration became especially urgent around a particular issue, temporary teams, made up of specialists from the various units involved, were formed. Managers in this organization also relied heavily on direct contact with their colleagues in other units. In this organization the managerial manpower devoted to integration was less than that in the plastics organization. Yet, compared with the container firm, the food organization was devoting a large amount of managerial time and effort to this activity.

Integration in the container organization was achieved primarily through the managerial hierarchy, with some reliance on direct contact among functional managers and on paperwork systems that helped to resolve the more routine scheduling question. Having little differentiation, this organization was able to achieve integration by relying largely on the formal chain of command. We are not implying that the other two organizations did not use this method at all. As Table VI–1 suggests, some integration did occur through the hierarchy as well as through paper systems in both of these organizations. But the great differences among functional managers seemed to necessitate the use of additional integrating devices in these two organizations.

From this discussion we can see another partial determinant of effective conflict resolution (in addition to those discussed in Chapters III and V) . This is the appropriateness of the choice that management makes about formal integrating devices. The comparison of these devices in these three high-performing organizations indicates that, if they are going to facilitate the process of conflict resolution, they should be fairly elaborate when the organization is highly differentiated and integration is thus more difficult. But when the units in the organization are not highly differentiated, simpler devices seem to work quite effectively. As we have already seen, however, the appropriate choice of an integrating device is not by itself sufficient to assure effective settlement of differences.

All of the plastics and food organizations, regardless of performance level, had some type of integrating device besides the managerial hierarchy. These devices were not equally helpful in interdepartmental decision making because, as we have pointed out, some of the organizations did not meet many of the other partial determinants of effective conflict resolution. However, there was evidence in all organizations that these devices did serve some useful purpose. To at least a minimal extent they helped to bridge the gap between highly differentiated functional departments. By contrast, in the low-performing container organization there was no evidence that the integrating unit was serving a useful purpose. Given the low differentiation within the organization, there seemed to be no necessity for an integrating department.

This comparison of the integrating devices in the three high-performing organizations points up the relationship between the types of integrating mechanisms and the other partial determinants of effective conflict resolution. We have stressed earlier that these determinants are interdependent. Even though we have not been able to trace the relationship systematically, this statement seems to include the final partial determinant, the choice of integrative devices. In all these organizations the choice of integrative devices clearly affected the level at which decisions were made as well as the relative influence of the various basic units.

We should also remember that any one of these determinants is only partial and that they should be seen as immediate determinants only. We have not explored the causes underlying them.

Comparison of Effective Conflict-Resolving Practices

Because of differences in the demands of each environment and the related differences in integrative devices, each of these high-performing organizations had developed some different procedures and practices for resolving interdepartmental conflict. However, certain important determinants of effective conflict resolution prevailed in all three organizations. We shall first examine the differences, then explore the similarities.

Differences in Conflict Resolution

The three effective companies differed in the relative influence of the various departments in reaching interdepartmental decisions. In the plastics organization it was the integrating department that had the highest influence. As we pointed out in Chapter III, this was consistent with the conditions in that organization's environment. The high degree of differentiation and the complexity of problems made it necessary for the members of the integrating unit to have a strong voice in interdepartmental decisions. Their great influence meant that they could work effectively among the specialist managers in resolving interdepartmental issues.

In the food organization the research and marketing units had the highest influence. This too was in line with environmental demands and with the type of integrating device employed. Since there was no integrating unit, the two departments dealing with the important market and scientific sectors of the environment needed high influence if they were effectively to resolve conflicts around issues of innovation. However, as we also indicated earlier, there was ample evidence that within these two units the individuals who were formally designated as integrators did have much influence on decisions.

The pattern of departmental influence in the container organization contributed to the effective resolution of conflict for similar reasons. Here the members of the sales and production departments had the highest influence. This was appropriate, since the top managers in these two departments had to settle differences over scheduling and customer service problems. If these managers or their subordinates had felt that the views of their departments were not being given adequate consideration, they would have been less effective in solving problems and implementing decisions.

Here again we have been restating comparatively the findings reported in earlier chapters. Such reiteration helps us to understand how this factor of relative departmental influence contributes to performance in different environments. Each high-performing organization had its own pattern, but each of these was consistent with the demands of the most critical competitive issue.

A second important difference among these three organizations in respect to conflict resolution lay in the pattern of total and hierarchical influence. The food and plastics organizations had higher total influence than their less effective competitors, and, related to this, the influence on decisions was distributed fairly evenly through several levels (Figure VI–1). The lower-level and middle-level managers who had the necessary detailed knowledge also had the influence necessary to make relevant decisions. In fact, they seemed to have as much influence on decisions as their top-level superiors. In the container industry, on the other hand, total influence in the high performer was lower than in the low performer, and the decision-making influence was significantly more concentrated at the upper management levels. This was consistent with the conditions in this environment. Since the information required to make decisions (especially the crucial scheduling decisions) was available at the top of the organization, it made sense for many decisions to be reached at this level, where the positional authority also resided.

The importance of the differences in these influence lines can be better understood if we let some of the managers in each organization speak for themselves. In the plastics organization lower and middle managers described their involvement in decisions in this way:

Figure VI–1

DISTRIBUTION OF INFLUENCE IN THREE HIGH-PERFORMING ORGANIZATIONS [a]

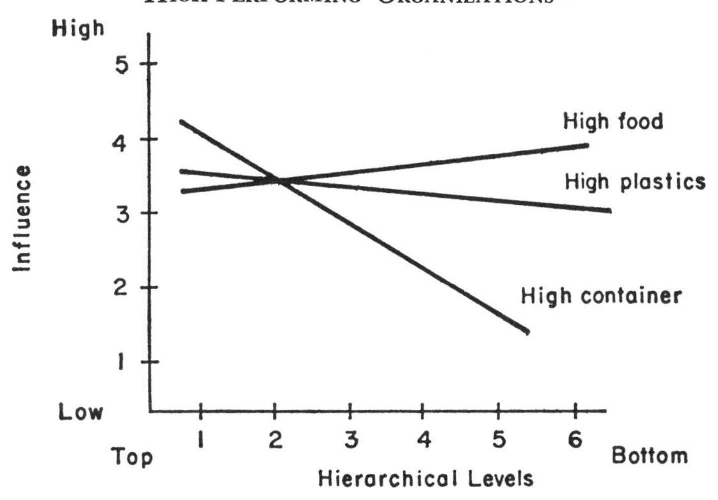

[a] Lines fitted by least square method. The difference in the slope of the lines between the high-performing food and the high-performing container organization was significant at .001. This difference between the high-performing plastics and the high-performing container organization was significant at .005. There was no significant difference between the food and plastics organizations.

When we have a disagreement ninety-nine times out of a hundred we argue it out and decide ourselves. We never go up above except in extreme cases.

* * * * *

We have disagreements, but they don't block progress, and they do get resolved by us. I would say on our team we have never had a problem which had to be taken up with somebody above us.

* * * * *

We could use these teams to buck it up to the higher management, but I think this would be a weak committee and a weak individual, and I am not willing to give my freedom up. They give you all the rope you need. If you need their help, they are there; if you don't need them, don't bother them.

The last manager quoted went on to substantiate a point made by many of his colleagues: While lower and middle managers made most decisions at their own level, they also recognized that major issues, which might have implications for products other than their own, should be discussed with higher management. But this discussion always took place *after* they had agreed on the best course of action for their own products.

Over and over, these lower and middle managers indicated their own responsibility for decisions and their feeling that to ask their superiors to resolve conflicts would be to acknowledge their own inadequacy. A higher-level manager stressed that this was also the view at his level:

> Top management has told these fellows, "We want you to decide what is best for your business, and we want you to run it. We don't want to tell you how to run it." We assume that nobody in the company knows as much about a business as the men on that team.

This same flavor was evident in remarks gathered in the food organization. Here, too, middle and lower managers stressed their own involvement in decisions.

Given these facts, the reader may be wondering about the activities of the upper echelons of management in the plastics and food organizations. If they were not involved in these decisions, what were they doing? While we made no detailed study of their activity, the data collected in interviews indicated clearly that they had plenty to keep them busy. First, they had the problems of administering their respective functional units. Second, they reviewed decisions made by their subordinates to make certain that the specialists working on one part of the product line were not doing anything that would adversely affect another part. In addition, in their dynamic environments they were constantly concerned with the search for new and longer range opportunities, which would fall outside the purview of any of their subordinates. In this regard we found in all the effective organizations the managers' time horizons became longer ranged as one moved up the hierarchy. This tendency was particularly marked in the plastics and food organizations. This, too, suggested that top executives in the food and plastics organizations were heavily involved in longer range issues and problems.

The tone of comments by managers in the container organization about who made decisions was dramatically different from that in the other two organizations. We cited some examples in Chapter V, but a few more at this juncture may emphasize the contrast. The middle and lower managers in the container organization emphasized the chief executive's and the other officers' roles in decision making:

> My primary contact is with [sales vice president and the chief executive]. This contact is around who we are going to give the containers to, because of our oversold position. They will determine which ones we are going to take care of. . . . Actually, what you really need though is [the chief executive's] decision. I usually start out these kinds of conflicts with [the production scheduling manager], but when somebody has to get heard, it ends up with [the chief executive]. Usually I am in contact with him three or four times a day.
>
> * * * * *
>
> When there is a problem I try to tell [production vice president] the facts and make some recommendations. He makes the decisions or takes it up to [the chief executive]. He doesn't get reversed very often. Sometimes he may say to me, "I agree with you, go ahead and do it," and then [the chief executive] will change it.

The sales vice president explained his own involvement, emphasizing application of the available facts:

> [The chief executive] holds a weekly scheduling meeting on Monday, which includes him, myself, the scheduling man-

ager, and a couple of the sales managers, depending upon what the crucial problems are. The scheduling manager has prepared the schedule on Friday. On Monday we tear it apart. This business is like playing an organ. You've got to hit the right keys, or it just doesn't sound right. The keys we play with are on the production schedule. In these meetings, though, the final decision rests with [the chief executive]. He gets the facts from us, and we influence the decision, but if there is any doubt, he decides.

All these comments serve to underline the differences in the distribution of influence between plastics and foods, on the one hand, and containers, on the other. These differences directly reflect differences in their respective environments.

Similarities in Conflict Resolution

So far, we have accentuated the important differences in these organizations in terms of the determinants of conflict resolution. Let us now look at some similarities. First, however, we should stress again that the differences actually stemmed from a fundamental similarity: Each of these organizations had developed conflict-resolving practices consistent with its environment.

The first major similarity among these organizations is in the basis of influence of the managers most centrally involved in achieving integration and resolving conflict. In all three organizations these managers, whatever their level, had reputations in the company for being highly competent and knowledgeable. Their large voice in interdepartmental decisions was seen as legitimate by other managers because of this competence. To return to the point made earlier, the positional influence of the managers assigned the task of helping to resolve interdepartmental conflict was consistent with their influence based on competence. Unlike the situation in some of the low-performing organizations, these two important sources of influence coincided in all these effective organizations. This point is illustrated by comments about the competence of the managers centrally concerned with conflict resolution in each organization.

In the container company, as we indicated earlier, the chief executive was regarded as extremely knowledgeable about the various facets of the business. As one manager expressed it:

> The fact is, as I understand it, that he is almost a legend in the industry. He knows every function in this company better than any of the people who are supposed to be handling that function.

But the chief executive was not the only one who had this respect. Managers in this organization also emphasized the knowledge and ability of the other top executives. A research engineer described the competence of the research director:

> I think another thing related to the close supervision I receive is the nature of the [research director]. He is an exceptional kind of guy, and he seems to know all the details and everything going on in the plant, and in the lab. He is continually amazing people in this regard.

A similar point was made about the production vice president by one of his plant managers:

> Oh yes, I hear from [the production vice president], but if he wants you, you are in trouble. You hear from him for sure, if your figures are too far off. He is pretty understanding. If you can explain, he understands. He also can really help you out on a serious production problem. He can tell you what to do. He knows just how far a job should be run before it should be pulled off.

In this organization, as these comments suggest, the knowledge and expertise of the top managers gained them respect from their subordinates and legitimated their strong influence over decisions. In the foods and plastics organizations the knowledge-based influence worked in a similar manner to justify the high influence of the middle managers centrally involved in helping to resolve interdepartmental conflict. Comments similar to those cited in earlier chapters may help to highlight this point. An integrator in the food organization explained the importance of expertise in his job:

> Generally, the way I solve these problems is through man-to-man contact. I think face-to-face contact is the very best thing. Also, what we [the integrators] find is that most people develop a heavy respect for expertise, and this is what we turn to when we need to work out an issue with the fellows in other departments.

Similarly, a fundamental research scientist in the plastics organization indicated (as did many others in this organization) that he believed the members of the integrating unit to be competent, which helped them to achieve collaboration:

> I believe we have a good setup in [the integrating unit]. They do an excellent job of bringing the industry problems back to somebody who can do something about them. They do an excellent job of taking the projects out and finding uses for them. In recent years I think it has been staffed with competent men.

In all three high-performing organizations, then, our data suggest a consistency in three factors that helped those primarily responsible for achieving integration to settle interdepartmental disputes. The managers who were assigned the responsibility for resolving conflict were at a level in the organization where they had the knowledge and information required to reach interdepartmental decisions and they were regarded as competent by their associates. Thus (1) *positional influence,* (2) *influence based on competence,* and (3) *the actual knowledge and information required to make decisions* all coincided. While there was this similarity, as we pointed out above, the level at which influence and knowledge were concentrated varied among the organizations because of differences in the certainty of their respective environments.

A second important similarity in these three organizations lay in the mode of behavior employed to resolve conflict. All three, as we have seen, relied heavily on open confrontation. The managers involved in settling conflicts were accustomed to open discussion of all related issues and to working through differences until they found what appeared to be an optimal solution. This was so regardless of the level at which the conflicts were handled. Typical comments from managers in each of the three organizations illustrate this point more vividly than the numerical data reported earlier. A researcher in the plastics organization described how he and his colleagues resolved conflicts:

> I haven't gotten into any disagreements yet where we let emotions stand in the way. We just go to the data and prove out which is right. If there is still some question about it, somebody can do the work to re-examine it. Emotions come up now and then. However, we usually have group decisions, so if I am not getting anywhere, I have to work it out with the others.

A production engineer in the food organization expressed a similar viewpoint:

> We often will disagree as to basic equipment. When we can't agree on what equipment to use, we will collaborate on some tests [with research], and sometimes we will run it both ways to find out what is the best way. Actually, the way this works out, one of their fellows and I will be at each other's desk doing a lot of scratching with a pencil trying to figure out the best answer and to support our point of view. We will finally agree on what is the best way to go. It is a decision we reach together.

The director of research in the container organization discussed his role in the resolution of conflict with the chief executive:

> I am sure a lot of people would say this is a one-man company. Sure, [the chief executive] keeps close tabs on the dollars, and I must keep good score for him in regard to everything we spend. He is pretty gentle with me, and I have no run-ins with him. He talked to me this morning about a problem, and I knew that regardless of whether I said yes or disagreed with him he would have gone along and taken my advice. He likes to complain a lot, and holler and bellow and be like a wild bull, but he gives up when he sees a good case. He'll ask for a real good story, and we have to give it to him, but if it *is* a good story, he will go along with us.

We should emphasize several important points about this comment. It and similar remarks from the major executives in the container organization indicated that while the chief executive was strong and dominant, he expected to have all points of view and pertinent information discussed before making a decision. These responses likewise indicated that there was give and take in these discussions and that the other major executives often influence the outcome, if the facts supported their point of view. It is also worth noting, as a comment from a plant manager in this organization suggests, that lower managers used the same method to resolve conflicts:

> I'm an easy-going sort of fellow, but I get mad sometimes. When we get something to fight about, we just say it, face the problem, and it is over. We get the issue out on the table and solve it. It has to be done that way. [The production vice president] does it that way. We all follow his lead.

While these statements all deal with technical issues, we could cite similar comments concerning marketing problems. The important fact to emphasize is that these three organizations relied on confrontation as a mode for resolving interdepartmental conflict to a greater extent than all but one of the other organizations (the low-performing food organization). This fact does not seem unrelated to the importance of competence and knowledge as a basis of influence for the managers primarily responsible for resolving conflicts. High value was traditionally placed on knowledge and expertise in all three organizations. Consequently, managers were very willing to see disagreements settled on this basis.

This reliance on confrontation suggests another important characteristic of all three organizations: Managers must have had sufficient trust in their colleagues and, particularly in the case of the container organization, in their superiors to discuss openly their own points of view as they related to the issues at hand. They seemed to feel no great concern that expressing disagreement with someone else's position (even a superior's) would be damaging to their careers. This feeling of trust apparently fostered effective problem solving and decision making.

Summary Comparison of the High-Performing Organizations

The plastics organization, which functioned in the most dynamic and diverse of the three environments, was consequently most highly differentiated of the three high-performing organizations. Since this condition could create major problems in maintaining the required state of integration, this organization, as we have seen, had developed an elaborate set of formal devices (both an integrating unit and cross-functional teams) to facilitate the resolution of conflict and the achievement of integration. Because market and scientific factors were uncertain and complex, the lower and middle echelons of management had to be involved in reaching joint departmental decisions; these managers were centrally involved in the resolution of conflict. This organization also met all the determinants of effective conflict resolution. The integrators had balanced orientations and felt that they were being rewarded for the total performance of their product group. Relative to the functional managers they had high influence, which was based on their competence and knowledge. In resolving conflict all the managers relied heavily on open confrontation.

In contrast to the plastics organization, the container organization was in a relatively stable and homogeneous environment. Thus its functional units were not highly differentiated, which meant that the only formal integrating device required was the managerial hierarchy. But in using this device this organization also met the determinants of effective conflict resolution. The sales and production units, which were centrally involved in the crucial decisions related to scheduling and delivery, both felt that they had much influence over decisions. Around these issues influence was concentrated at the top of the organization, where top managers could centrally collect the relevant information to reach decisions. Middle managers, particularly those dealing with technical matters, did have some influence. The great influence of the top managers stemmed not only from their position, but also from their competence and knowledge. Finally, conflicts between departments were resolved and decisions reached through problem-solving behavior.

In these two paragraphs we have described two quite different organizations, each of which is well equipped to deal with its own external environment. Another way to understand the contrasts between them is to examine the major sources of satisfaction and of stress for the executives in each. While we made no systematic effort to collect such data in the plastics organization, the contrast between the two organizations can be clearly seen from interview comments of the managers in each organization. Managers in both organizations were generally quite well satisfied with their situations, but they were finding satisfaction for some quite different reasons. In the plastics organization an important source of satisfaction was the active involvement in decisions. Middle managers often expressed the feeling that they were running their own firms. One product manager in the sales department put it this way:

> Our present organization allows us as individuals to more formally play a role in decision making, which we didn't do before. Now, with the teams, we can make a decision which will affect the profit. We can see the results of our efforts more realistically than we could before. Now that it has management approval, it has a nice flavor. It's nice to be doing something they approve of. The product manager has no formal authority. But putting him on the team gives him some sort of authority. I'm not sure what kind of authority it is, but it makes my job more meaningful. . . . Of course, we all recognize that the other guys on the team are depending upon our effort, so we make an effort to produce.

Managers in the container organization, however, indicated that they liked their jobs for quite a different reason—because they knew where to get a decision made. One manager expressed it in this manner:

> He [the chief executive] does all the scheduling himself, and in essence what you have is a large organization run by one man. This is a refreshing switch from the organization where I had previously worked. I find this very beneficial. If I want something decided, I can go right to him and get a direct decision. You tell him what you want to do, and he will tell you right then and there whether he will let you do it or whether he won't.

The sources of dissatisfaction and stress in the two organizations were also different. A manager in the plastics organization described some of the points of concern to him:

> I worked for another company which was different, where there were fairly definite lines of authority. This place was quite a revelation to me. In my old company we always knew whose jobs things were. Occasionally here we run into situations where we don't know whose jobs things are. . . . All of these meetings take a lot of time. I used to spend eight hours by myself, and I thought I could get more things done. I feel now that I spend time on committees instead of making autocratic decisions, but this isn't really a disadvantage, as we do get better solutions. . . . Also, there can be conflict between your position as a functional manager and as a team member. The more empathy with others you have, the worse it gets.

What disturbed this manager and a few others was the ambiguity of responsibility and relationships in this organization. Many managers often had dual loyalties—to their functional superiors and to their team colleagues. They had to decide themselves what needed to be done. The involvement of many managers in interdepartmental decision making made these difficulties unavoidable, and it also meant that managers who had a low tolerance for ambiguity and uncertainty did not always enjoy their work.

In contrast, the few managers in the container organization who expressed dissatisfaction were most concerned because upper managers seemed to be so involved in their activities. As one man said:

> Your boss is telling you to check something, and then he jumps down your throat five minutes later. They should know what you are doing and try to give you some answers, or else they should let you do it. . . . I know this job involves a lot of pressure, particularly because at first you are just getting ignored around here and then they are jumping on you, and the pressure is really acute. Somebody has to be the whipping boy around here, and that is just part of this job.

These data suggest two things. The first is quite obvious—that these two organizations were quite different places in which to work. The second inference is more speculative. There is some suggestion, from the tone of the interviews, that the managers in the two organizations had somewhat different personality needs. Those in the plastics organization seemed to prefer more independence and had a greater tolerance for ambiguity, while those in the container company were perhaps better satisfied with greater dependence upon authority and were more bothered by ambiguity. While there may have been these differences in personality needs, each organization (as well as the food organization) seemed to provide a setting in which many members could gain a sense of competence in their jobs. As we suggested in Chapter I, this provided them with important sources of satisfaction. The fact that so few managers in either organization did express any dissatisfaction with such different organizational climates would suggest that this is so. While we have no way to confirm this speculation, it does raise again the importance of the point made earlier, that the organization must fit not only the demands of the environment, but also the needs of its members.

In any case, the contrast between the plastics and the container organizations is very sharp. In a sense, they represent opposite ends on a continuum, one dealing with a very dynamic and diverse environment, where innovation is the dominant issue, while the other is dealing with a very stable and homogeneous environment, where regularity and consistency of operations were important. The food organization, as our discussion has suggested, was in many ways like the plastics organization. The differences between them seemed to be more of degree than of kind. While the food environment was not so dynamic and diverse as that of plastics, it seemed to be toward that end of the continuum. The integrating devices, although not so elaborate as those in the plastics organization, were of the same nature, designed to provide linkage at the middle and lower managerial levels. The two organizations met most of the same determinants of effective conflict resolution. The major difference between them was that the plastics organization appeared to be devoting more of its managerial manpower to devices that facilitated the resolution of conflict. The important point, however, is that the food organization, like the other effective organizations, had developed a set of internal states and characteristics consistent with the demands of its particular environment.

We should, however, recognize one limit to this conclusion. Each of these organizations had developed characteristics that were in tune with the demands of its *present* environment. Whether these same characteristics will provide long-run viability depends, of course, on whether the environmental demands change in the future. Given the widely observed tendency toward greater scientific, technological, and market change, the plastics and food organizations would seem to be in a more favorable position to maintain their high performance. Major technological or market changes in the container industry would almost certainly create serious problems for the high-performing container organization. This suggests that the managements in stable industries must develop within their organizations some capabilities for watching for environmental changes and preparing to adapt to them. It also suggests that in the future more and more organizations may resemble the high-performing plastics and food organizations.

A Contingency Theory of Organizations

From this comparison we have seen that it is possible to understand the differences in the internal states and processes of these three effective organizations on the basis of the differences in their external environments. This, along with the comparison between the high performers and the other organizations in each environment, has provided us with some important leads as to what characteristics organizations must have in order to cope effectively with different environmental demands. These findings suggest a contingency theory of organization which recognizes their systemic nature. The basic assumption underlying such a theory, which the findings of this study strongly support, is that organizational variables are in a complex interrelationship with one another and with conditions in the environment.

In this study we have found an important relationship among external variables (the certainty and diversity of the environment, and the strategic environmental issue), internal states of differentiation and integration, and the process of conflict resolution. If an organization's internal states and processes are consistent with external demands, the findings of this study suggest that it will be effective in dealing with its environment.

More specifically, we have found that the state of differentiation in the effective organization was consistent with the diversity of the parts of the environment, while the state of integration achieved was consistent with the environmental demand for interdependence. But our findings have also indicated that the states of differentiation and integration are inversely related. The more differentiated an organization, the more difficult it is to achieve integration. To overcome this problem, the effective organization has integrating devices consistent with the diversity of the environment. The more diverse the environment, and the more differentiated the organization, the more elaborate the integrating devices.

The process of conflict resolution in the effective organization is also related to these organizational and environmental variables. The locus of influence to resolve conflict is at a level where the required knowledge about the environment is available. The more unpredictable and uncertain the parts of the environment, the lower in the organizational hierarchy this tends to be. Similarly, the relative influence of the various functional departments varies, depending on which of them is vitally involved in the dominant issues posed by the environment. These are the ways in which the determinants of effective conflict resolution are contingent on variations in the environment. Four other determinants, however, seem to be interrelated only with other organizational variables and are present in effective organizations in all environments. Two of these are the confrontation of conflict and influence based on competence and expertise. The other two factors are only present in those effective organizations that have established special integrating roles outside the managerial hierarchy—a balanced orientation for the integrators and a feeling on their part that they are rewarded for achieving an effectively unified effort. Our findings indicate that when an organization meets most of these determinants of effective conflict resolution, both the general ones and those specific to its environment, it will be able to maintain the required states of differentiation and integration.

This contingency theory of organizations suggests the major relationships that managers should think about as they design and plan organizations to deal with specific environmental conditions. It clearly indicates that managers can no longer be concerned about the one best way to organize. Rather, as we shall see, this contingency theory, as supported and supplemented by the findings of other recent research studies, provides at least the beginning of a conceptual framework with which to design organizations according to the tasks they are trying to perform. We shall now examine in more detail the implications of these findings, not only for the design and planning of organizations, but also for gaining a clearer perspective on current organizational theory.

CHAPTER VII

Traditional Organizational Theories

ALL THE EVIDENCE generated by the line of research that this book reports has now been presented. We have analyzed this evidence and drawn conclusions from it with different degrees of certainty. At this point some books would end. Our purpose, however, is not only to report on research but, equally importantly, to use this research to throw new light on some of the current contradictions and confusions in organization theory. We alluded to this state of the theory in Chapter I, but the sources of confusion will become increasingly apparent in this chapter and the next, as we review the theory in somewhat more detail.

As we proceed, we will be asking a new type of question. Rather than asking, What are the data, and what do they mean? we will now ask, Can the findings of this study give greater unity and generality to the pre-existing knowledge about organizations? This question needs exploring, though the task is an ambitious one.

But how to proceed? We will need some rationale for selecting among theories, since we have too many choices rather than too few. As practicing social scientists, the authors are tempted to approach the problem in terms of the history of sociological thought on organizational issues. This would draw us into reviewing such important contributions as those of Weber, Parsons, Merton, Homans, and Selznick. But since this book is addressed primarily to practicing administrators, such a course would involve a language which, for all its merits, would create great difficulties for many readers.

We could swing to the other extreme and look at the way new ideas about organizational practice usually impinge on the administrator. The last 50 years of executive life have been filled with a multitude of important new methods and techniques for running a business. Just a partial listing of the major methods is revealing. Around the turn of the century business was being urged to systematize in terms of cost accounting, production control, and budgeting. Then the scientific management movement came in, with time and motion studies and job- and work-flow rationalization. About the same time a plethora of monetary incentive systems, such as the infamous Bedaux plan, came into vogue. During the thirties and the war period the human relations approach sprang up, with its emphasis on upward communication, listening, and participative management. Since the war we have seen a real explosion. Consider the following list of some of the better-known techniques:

1. PERT and critical path systems
2. Business gaming
3. Value analysis
4. Sensitivity or T-group laboratories
5. Long-range planning techniques
6. Decentralization, profit-center management, unit management
7. System design techniques
8. Creativity training—brainstorming or synectics
9. Operations research, with linear programming, dynamic programming, etc.
10. Management grid training
11. Cost effectiveness analysis
12. Scanlon plan and other profit-sharing techniques
13. Decision theory—decision trees, allocation models, etc.
14. Motivation laboratories
15. Human factors engineering

This list, though far from complete, does suggest the multitude of voices directed at the administrator, each clamoring for the legitimacy of its particular way of improving organizational effectiveness. Trying to review this welter of techniques one at a time seems hopeless. For our purposes, it is too close to where the practitioner lives.

Before we leave this list, it is useful to note one particular way of sorting it. Each of these techniques seems to carry with it a thrust in one of two directions—either toward greater order, systematization, routinization, and predictability, or toward greater openness, sharing, creativity, and individual initiative. One thrust is to tighten the organization; the other, to loosen it up. The authors' own rough attempt at ordering these techniques is to put all the odd-numbered items in the first camp and the even in the second. What is going on here? Are opposite forces that simply offset each other building up in modern organizations—and pity the manager who is caught between them? All these techniques must have something to offer, or they would not receive so much support as they have. It is unlikely that one set is all right and the other all wrong. We will come back to this question later.

Since this book is addressed to administrators, we must consider approaching our review of organization theory in terms of the language managers themselves currently use. One set of widely practiced ideas is conveyed in such expressions as "authority," "responsibility," "line and staff," "functional vs. product organization," "chain of command," etc. The theorists who emphasize this kind of language trace their intellectual heritage back to such men as Fayol, Mooney, Urwick, Gracunias, and Gulick. These writers have come to be known as the "classical" or "administrative process" school. This school has been subjected to a great deal of criticism in the past 20 years. The fact that its language persists in business usage, however, is sufficient justification for a fresh look at this theory in the light of the findings of our study.

Another language widely used by administrators centers around the word "relationship." Managers are regularly heard talking about establishing relationships or contacts with sundry people—often outside their official jurisdiction. They refer to the "state" of different relationships using such adjectives as "cordial" or "strained," "straightforward" or "confused," "open" or "sticky," "warm" or "distant." They spend time and energy to develop important relationships, both between individuals and between groups. Into the building of interpersonal relations the concept of personality, as well as various concepts having to do with communication, is often introduced. Relations within and among groups also receive constant attention. Of course these concepts have been emphasized by the human relations theorists, who are intellectual descendants of such men as Mayo, Roethlisberger, Lewin, and McGregor. These pioneers and their successors have stressed the importance of viewing organizations as systems of relationships. Since we, the authors, have ourselves been brought up in this school, it will be especially necessary for us to be aware of its limits as well as its strengths as we examine it in the light of our present findings.

The two fields we have chosen—the classical and the human relations—certainly do not constitute all the current schools of thought on organization or management theory. Two other identifiable theories, for example, are the "decision theory" school, which has developed around the work of March and Simon and Cyert and March, and the "bureaucratic" school, which is based on the work of Weber but has more recently been clarified by Blau, Gouldner, Crozier, and others.[1] The two schools we will deal with do, however, seem to be most widely used by practitioners. Any more adequate or useful theory must build on or displace them. They have the further advantage of representing a middle level of abstraction that we can later connect to general sociological theory, on the one hand, and to the practical affairs of managers, on the other.

Finally, these two schools of thought have had a rather unusual historical relationship to each other. For roughly three decades these two approaches to organization theory have persisted side by side among managers and in schools of business, in spite of the fact that their implications for management action are at many points in direct conflict. Neither theory has been able to displace the other. Do our present findings illumine this curious relationship? What explains this hostile complementarity?

Classical Theory

The classical writers usually start their reasoning by using examples of very primitive organizations. One favorite, for instance, involves a man who wants to move a stone that is too heavy for him. So, of course, he arranges to secure the temporary services of a second or third man by offering a reward. When one pushes while another pulls, we have the beginning of a division of labor. When the first calls out the signal for a big heave, we have a primitive chain of command serving to integrate the differentiated parts of the system. From such simple examples evolves much of the final theory. They provide the rational basis for deducing many of the so-called "principles" that classical theorists have stressed.

Urwick describes one of these rules:

> The considerations which appeared of greatest importance were that there should be clear lines of authority running from the top into every corner of the undertaking and that the responsibility of subordinates exercising delegated authority should be precisely defined.[2]

We see here a prescription for three particular structural attributes: (1) limited and prescribed communication channels; (2) detailed role descriptions; and (3) authoritative leadership styles. Another structural attribute urged by Urwick is a narrow span of control:

> Students of administration have long recognized that, in practice, no human brain should attempt to supervise directly more than five, or at the most, six other individuals whose work is interrelated.[3]

Mooney supports these points explicitly or implicitly in saying:

> The type of management which regards the exact definition of every job and every function, in its relation to other jobs and functions, as of first importance, may sometimes appear excessively formalistic, but in its results it is justified by all practical experience. It is in fact a necessary condition of true efficiency in all forms of collective and organized human effort.[4]

Comments of this kind can be multiplied at length from these and other authors. There is no doubt that these men are saying that, in the terms of our study, a highly formalized structure with a directive or authoritarian leadership style will lead to high performance. Furthermore, as Mooney indicates, these writers see this type of structure as universally applicable. Urwick also makes this point in no uncertain terms:

> It is the general thesis of this paper that there are principles which can be arrived at inductively from the study of human experience of organization, which should govern arrangements for human associations of any kind. These principles can be studied as a technical question, irrespective of the purpose of the enterprise, the personnel composing it, or any constitutional, political or social theory underlying its creation.[5]

Since these writers saw no need for different kinds of organizations, it follows that they were not concerned with the sort of organizational differentiation that we have studied. To them differentiation was in terms of technical and physical facts *per se*, such as technical process differences, or product or geographical differences. They tended to assume that the greater the specialization in terms of these differences, the greater the concentration of expertise, and the better the organizational performance. They generally minimized the problems of integrating these different parts since, according to them, the very process of creating the division of labor into highly specified roles also created the chain of command that assured that the role performances would be carried out as specified. Thus the chain of command is *the* mechanism of integration, and its effectiveness can, by and large, be taken for granted.

How do we reconcile this theory with the findings of the study? It is revealing to note that these writers pick as starting examples organizations designed to perform such simple and obvious tasks as moving stones. The goal of moving a stone is about as clear-cut and certain as one can find. There will be no doubt when the task is completed and a minimum of uncertainty about the choice of methods (presuming, as these theorists did, that no appropriate machinery is immediately available). Given this kind of environment, the conclusion that the highly structured system they recommend will lead to high performance is consistent with the findings of our research. But obviously these men did not learn how to build organizations by rolling stones. We can throw a clearer light on this theory by examining the personal backgrounds of these writers, all of whom had considerable firsthand experience in organizations.

Urwick's early organizational experience was of two kinds. The first was as a young Oxford graduate in the British Army during World War I. Judging from his frequent references to it, this experience was apparently very important in shaping his ideas about organizations. But let us not jump hastily to stereotypes about military organizations. Not all military tasks are the same, and differences in goals tend to induce different kinds of organizations. Nevertheless, the massive armies arrayed against each other in the trench warfare of World War I were faced with a kind of task that must have generated tremendous pressure toward a highly structured system. The overall job of attacking or holding was clear and simple. The coordination of all details necessitated centralized planning. A very strong chain of command was required to carry out the innumerable details involved in the proper timing of widespread and diverse activities, to say nothing of the problem of maintaining discipline under such inhuman conditions. Urwick's other firsthand experience was in the production side of a chocolate candy company during the 1920s. Here again we can speculate that a highly structured organization was probably quite relevant in the context of this firm's environment.

Fayol, the senior writer in the classical school, had an extended career in the top management of a large coal mining company in France. It was in this setting that he developed his ideas about administration. Here again, we can speculate about the nature of his firsthand experience in a large organization designed to cope with an almost uniquely stable technological and market environment.

Similarly, Mooney, a vice president and director of General Motors, seems to have acquired his principal organizational experience in the automotive industry. It was as an officer of General Motors in 1931 that he wrote, with Allan Reiley, his principal contribution to the literature, *Onward Industry!* [6] While this was at the end of a decade of great growth in the automotive industry, it was nevertheless a relatively stable period in terms of technical change and even in terms of the nature of the consumer's demand. And, of course, the automobile assembly line has always been cited as the extreme example of a routinized technical process. Is it any wonder that out of this experience Mooney wrote of the advantages of a highly structured organization?

Knowing something of the personal backgrounds of some of the key contributors to classical organization theory helps us to understand why they strongly recommended a particular kind of organization model—a highly structured, authoritarian system. It does not, however, completely explain why these men made such dogmatic statements about the universality of this model and thereby exposed themselves to a lot of subsequent criticism. After all, they had some experiences broader than those we have cited. Fayol became involved in governmental organizations; Urwick, in international organizations; and Mooney, in consulting organizations. Again, some historical perspective makes their positions more understandable.

The quotations we have cited were written before the mid-1930s. At that time very few writers were even suggesting alternative ways of looking at formal organizations. About the only exceptions were Mary Parker Follett and Elton Mayo, in his early work. The general beliefs and mores of the times supported the classical position. The ideas were implicit in the legal definition of the corporation. They were rooted in the traditional concept of the master-servant relationship, which had been carried over to the employer-employee relationship. While the union movement was questioning this basis, the assumptions were still well established. It was in this historical context, which now seems so distant, that these authors felt justified in making such categorical statements about the one best way to run any kind of organization for any kind of purpose. Even today this idea, while publicly denied, persists in the practice of many administrators.

At this point we make special mention of a well-known writer of the classical school who was a notable exception to their "universal" approach. This was Luther Gulick. Many of his generalizations are in sharp contrast to those quoted above. Note the following:

> Students of administration have long sought a single principle of effective departmentalization just as alchemists sought the philosopher's stone. But they have sought in vain. There is apparently no one most effective system of departmentalism.[7]
>
> . . . organization must conform to the functions performed.[8]
>
> In the discussion thus far it has been assumed that the normal method of interdepartmental co-ordination is hierarchical in its operation. . . . In actual practice, there are also other means of interdepartmental co-ordination which must be regarded as part of the organization as such. Among these must be included planning boards and committees, interdepartmental committees, co-ordinators, and officially arranged regional meetings, etc.[9]

These statements certainly convey a tone that is much more in line with the findings of this research. In the light of Gulick's background, this is not so surprising. First of all, he was an academic, with a firsthand knowledge of the collegiate organization of the university. Second, his professional work was focused on governmental organizations, particularly on the myriad forms that appear in a large city government. Most of the examples he cites are drawn from these settings. How could one be a universalist after studying such a mixed bag of organizations, ranging from the police force to the city council?

We do not want our brief discussion of its origins to leave the reader with the impression that classical organizational theory has no current value. We are arguing instead that it presents a special case that can now be subsumed within a more general theoretical framework. The "principles" of the classicists can be treated as one end of a continuum. In fact, without even too much change, their key concepts can be translated into the conceptual framework of this research. As examples, let us discuss the scalar principle, the line-staff distinction, departmentalization, and finally, the related concepts of authority and responsibility.

The "scalar principle," as used by classical writers, is a complex or compound concept that we must break up into parts to relate to the concepts of this study. First, it concerns the process "by which this [supreme] coordinating authority operates from the top throughout the entire structure of the organized body. . . . In organizations it means the graduation of duties . . . according to the degrees of authority and corresponding responsibility." [10] This seems to refer to what we have described as the distribution of influence. The classical writers assumed that theirs was the only conceivable way to allocate authority in an organization—high influence at the top, with less at each descending level, to low or no influence at the bottom. In this study we have made no arbitrary prior assumption about the "right" way to distribute power and have treated influence distribution as a variable. A second idea buried in the scalar concept is that of coordination. Classicists assume that the superior coordinates the efforts of his subordinates. All conflicts, according to this theory, are referred to the shared superior for resolution. In this study we have seen that this is but one of various methods organizations actually use in settling differences and achieving integration. As we see it, the locus of conflict resolution is also a variable. It does not always reside in the boss, but can be built into a number of roles. Finally, the scalar concept suggests that each role in the hierarchy should be "precisely defined" in order to get clear lines of authority running from the top. This is part of what we have chosen to call degree of formal structure. The classicists tended to see high structure as always contributing to high performance. Once again, we have seen high structure as but one end of a continuum.

The classical writers gave a good deal of attention to the distinction between line and staff roles. In spite of this concentrated scrutiny, they failed to agree on any single useful definition. They were clear about the nature of the line role: Its occupant was the man in the chain of command who received delegated authority and responsibility from above and, in turn, passed it on to his subordinates and, by this process, coordinated their work. It follows from this definition that staff roles have no authority and are unnecessary to achieve coordination. This is what creates the problem. The factual existence of non-line executives doing coordinating work could not be denied, but the classical definition of the line role makes them all redundant. Classical writers resorted to lengthy circumlocution to avoid this logical fallacy—without much success. In this study no such dilemma arises. The line-staff distinction has no meaning in our terms. We do distinguish between integrating and specialist functions that exercise a mix of both knowledge-based and position-based influence. Since we never assumed that "line" roles do all the integrating or have all the influence, this comes as no shock.

The issue of departmentalization or division of labor was another central concern of the classical writers. They were searching for some central answer to the question of how to group sub-tasks so that the overall task of any organization could best be performed. They were chiefly preoccupied here with the issues of alternative ways of arranging spatial and technical differences among sub-units. Gulick is the only writer who gave significant attention to the differences in orientation that develop among sub-groups and the effects of these differences on the state of integration. Gulick developed some intriguing ideas, but he was not able to test them empirically. Of course the present research started off with this same concern; by tying differentiation to the environmental task and by demonstrating its antagonism to integration, we have made this issue amenable to systematic research. More detail on the practical implications of the findings for organizational design will be offered in the final chapter. Suffice it to say here that the usual structural choice between organizing sub-units along functional or product lines presents a trade-off, in our terms, between emphasizing differentiation at the expense of integration (functional sub-units), or *vice versa* (product sub-units). Except for Gulick the classical writers generally emphasized the positive effects of more and more specialization (differentiation) and assumed that the chain of command could handle the associated problems of integration. Once again, in the light of this study, we can see that the classical emphasis on the ultimate in specialization is but one end of a spectrum of choices that vary in their relation to performance, depending on the relevant environmental conditions.

The two most central and most elusive concepts in classical theory are those of authority and responsibility. The formal definitions of these terms are hard to find in the literature and are of limited use for our purposes. The concepts are so basic in the theory that their meaning has to be derived from the context in which they are used. Mooney's writing is of special help in this regard. Mooney finds he cannot discuss the distribution of authority in an organization without first introducing the idea of delegation. He writes: "The principle of delegation is the center of all processes in formal organization. Delegation means the conferring of a certain specified authority by a higher authority." [11] Elsewhere he refers to the "supreme coordinating authority." [12] Thus we see him using the term authority as if it can best be conceived as a finite substance residing initially with the supreme commander, who then parcels it out in specified amounts to others. This idea of authority has much in common with, and probably historically derives from, the legal concept of the rights of property, which include the right to appoint (and remove) agents to act in certain specified ways on the owner's behalf.

When Mooney gets to the very bottom of the command hierarchy, he sees clearly that the workers have authority to act in regard to the material resources of the organization. "Even the foreman of the section gang delegates an authority to his men, an authority to do certain things."[13] But Mooney recognizes that he is getting into deeper water when he extends this notion of authority for things to authority over people:

> The answer [to getting the job done] is not found merely in the exercise of his authority; the right of command to tell one person to do this and another to do that. . . . The exercise of leadership presupposes, as its first necessity, the power of understanding. The true leader must know in its entirety what is intended and know it so clearly that he can see the end from the beginning. Such knowledge is a necessity of leadership.[14]

If we remove its pomposity, this statement reduces to the argument that, to be effective, a manager's actions toward others must be based on the best available knowledge of the situation at hand—a contention with which few would quarrel and which the empirical findings of this study support. We would, however, be missing the important point if we ignored the aspect of Mooney's statement that strikes us as pompous: that for his kind of organization to work effectively, people need to *believe* that the boss is a "true leader"—one who clearly sees all things from beginning to end and therefore is a proper person to obey without question. This is the *assumption* that superior authorities have superior knowledge.

In our study the concept of influence has been important, and we intend it to subsume the classical notion of authority. The classical authority based on delegation of power we call position-based influence. This gives the incumbent of a specified position certain powers to commit organization resources, based on the delegated legal rights of property, and to influence the behavior of others, based on some control of the rewards and punishments the organization can bestow. We reject the classical assumption that knowledge automatically coincides with position-based influence. Rather, we see knowledge-based influence as a variable that it is important to keep separate. We would define knowledge-based influence as the ability of certain persons *as persons* to influence the commitment of organization resources and to influence the behavior of others, based on the relevance of the information available to them, the soundness of their reasoning, and their reputation for being right in the past. This is an entirely different foundation for influence from position, and both are important organizational variables.

The classical theory also tended to assume that authority is unitary and homogeneous. You either have authority, or you have not. We see the matter differently. Since a manager can have relatively high influence on one class of organization decisions and relatively low influence on another, it follows that influence on different kinds of issues will be distributed throughout the organization according to different patterns. While we have not systematically tracked down all such issues in the organizations we have studied, we consistently picked the most critical class of issue around which to study the influence pattern. In plastics and food it was innovation, and in containers it was customer service. Many other types of decisions were, of course, being made in these organizations, and we would expect a study of them to reveal varied patterns of distribution for both position-based and knowledge-based influence.

Classical theory also treated authority as a measurable and fixed commodity. The chief executive, by this reasoning, had 100% of the authority and, obviously, could parcel out to his subordinates no more than he initially started with. We see the concept of influence differently. The total amount of influence in an organizational system is not fixed but expansible, in much the same way as credit in a banking system. This can be true because influence is essentially a matter of the perceptions that arise from transactions among people. If these transactions are heavily constricted and limited, the people *in toto* will see themselves as having very little influence on decisions, and we can accurately say that the system has a comparatively limited total amount of influence. The opposite is true if the transactions are many and meaningful: The members of such an organization will see themselves as having considerable influence on decisions. Thus, in our model the total amount of influence in a system is not fixed but a variable.

One final point of difference must be made. We have seen that implicit in the classical writing was the need to maintain the belief that people with authority (positional influence) also had the relevant knowledge. If managers were to sanction such a belief, they would either have to believe it themselves, or at least act toward their positional subordinates as if they believed it. This style of managerial behavior is what we have termed as directive or authoritarian. Again we conceive of this not as the only managerial style, but as one point in a continuum of styles.

On one fundamental point the present research and the model that underlies it is in agreement with the classical handling of the authority concept. Mooney implicitly supports our finding that the quality of decision making is improved when positional influence is assigned to people with knowledge-based influence. This can equally well be stated the other way around—that the quality of decisions will improve when people with position-based influence also achieve knowledge-based influence.

The classical concept of responsibility is at least as elusive as the notion of authority. The typical statement is that the delegation of authority carries with it a corresponding responsibilty for doing what is authorized. In that sense the term responsibilty connotes such ideas as commitment, concern with results, and willingness to accept blame or praise for outcomes. It further suggests that when one is given positional influence, one should (if the system is to work well) adopt such an attitude of concern. Of course no one would be likely to argue with the proposition that it is helpful to system performance for those exercising positional influence to care about results. But the rub comes when the classical writers assume that this attitude automatically follows from the assignment of positional influence. In Mooney's words:

> The subordinate is always responsible to his immediate superior for doing his job, the superior remains responsible for getting it done, and this same relationship, based on coordinated responsibility, is reported up to the top leader, whose authority makes him responsible for the whole.[15]

The last phrase—that authority makes him feel responsible—states the assumption most clearly. This, of course, avoids the whole problem of motivation. The history of industrial organizations indicates that managers cannot afford to duck this question.

It is an exaggeration, however, to say, as some do, that the classical writers completely ignored the issue of motivation. They simply dealt with it in a limited way, using other terms. They did not want to seem crass, but every manager knows that buried in the statement, "the subordinate is always responsible to his immediate superior for doing the job," is the simple promise that while you might get rewarded for doing the job well, you will for sure get punished if you do not. In other words, the classical writers knew all about the rewards and punishments the organization can dispense through its chain of command. In fact, they seem to have relied on them completely to induce the desired sense of responsibility. In so doing they ignored the rewards and punishments from peer groups. They also ignored the other major process by which a sense of responsibility is induced. This is the gradual process whereby the individual slowly comes to so identify his own ego with his job role that he is pushed from inside to care about performance in order to satisfy his need to demonstrate competence and mastery.

The classicists' inadequate treatment of the entire issue of motivation provides a natural transition into our consideration of the human relations school, since historically it was this lapse in classical theory that stimulated the rise of the newer theory. In retrospect, we see that we have been able to use the concepts and findings of our study to place in a fresh perspective the classical approach to organization theory. The concerns and biases of the writers are understandable on the basis of their own individual experiences, and their "principles" are roughly translatable into the variables of this research. The essential point to emphasize is that their principles have only limited applicability, since they meet only the organizational demands generated by relatively stable and homogeneous environments.

Human Relations Theory

The human relations theorists approached the study of organizations by calling attention to "the seamy side of progress," to use Mayo's apt phrase. The early writers in the 1930s were concerned with the many signs that modern industrial organizations were generating some undesirable human consequences as well as a vast flow of goods and services. They saw these signs in labor-management conflict, in worker apathy and boredom, in endless struggles for power among managers—all pointing to a large-scale waste of human resources. From this beginning the research of the human relations school has grown over the years to encompass an interest in individual and small-group behavior, intergroup behavior, and total organization phenomena. In reviewing it we will for several reasons draw our examples primarily from Roethlisberger's and Dickson's original studies at the Hawthorne Works of Western Electric.[16] In the range of topics considered these studies foretold all the major lines of subsequent inquiry. They also attracted widespread interest among managers, which did more than any other single study to make the human relations approach a key part of most managers' everyday theory of organization. This means not only that the studies will be familiar to many readers but also that they will help us to understand how this line of research started a movement to change management practices. Some of the many other important contributions to the theory that have come since will be mentioned in passing.

The study of the bank wiring room at Hawthorne pointed up the discrepancy between how the organizational system was supposed to work and how workers actually behaved. It dramatically demonstrated how informal work groups provide mutual support and effective resistance to management schemes for increasing output. In spite of management's proper application of classical organization theory, workers were not acting "responsibly." They were not motivated to perform their assigned tasks to the best of their ability. They were more interested in the rewards and punishments of their work group than in those of management. At this stage the problems posed by the inadequacies of classical theory were stated—but no answers or remedies were apparent.

This highlighting of the great gap between the world of management and the world of workers was related to an early interest in the problems of communications up and down the organization hierarchy. The relay assembly test room was the stage for an early experiment in communication between workers and management. The favorable output results of this experiment attracted considerable attention. It probably encouraged some writers to give blanket endorsement to the idea of participative management as a way of increasing the workers' involvement in and motivation toward achieving the goals of the enterprise.

Another lead came from the development at Western Electric of individual counseling, with its attention to interpersonal processes. This program highlighted the potential for learning and problem-solving in two-person relationships when one person listens skillfully as the other explores and thinks through his concerns and feelings. The parallel development by Carl Rogers and his associates of the nondirective therapy methods greatly enhanced the interest.[17]

After the publication of the Western Electric studies, the pace of human relations research accelerated. Another early experiment that may have given even greater impetus to generalizations about participative management was conducted by Lewin's associates, White and Lippitt.* This work has since been complemented by attention to the opportunities for listening and learning in group settings. More recently, considerable attention has been given to the problems of communications between groups in organizations.[18] All of this led to identification of a need for upgrading the interpersonal competence of managers in order that they might cope more effectively with these myriad communication problems.[19] Emphasis was placed on developing trust and openness in all organizational relationships.[20]

These historical developments in the human relations approach have been summarized only briefly, because many recent books on management have dealt with them, and they are generally well known. This is in contrast to the classical theory, which, while widely practiced, has received less explicit treatment in the current literature. Altogether, these

* This semicontrolled experiment exposed several boys' groups to different styles of leadership that were characterized as autocratic, democratic, and laissez-faire. Very careful records were kept on the behavior of the boys and their leaders. The different styles of leadership and responses to them were spelled out in specific operational terms. The general finding that caught the public's attention, however, was that the democratically run groups were more effective in that they were at least as productive as any and also more creative.

human relations ideas have not only added a good deal to our knowledge about human behavior in organizations, but have also created a pressure on management to change the more customary ways of running organizations. It is this normative and prescriptive aspect of this work that we especially want to compare with the classical approach and to review in the light of the present study. The researchers and writers identified with human relations in industry have by no means all agreed on these matters, and many have avoided prescribing courses of action to managers. In spite of this, the human relations approach has acquired a certain popular image, the net effect of which has been to push managers toward:

(1) Securing the participation of lower echelons in solving the organization's problems, and
(2) Fostering more openness and trust among individuals and groups in organizations.

It is in this form that the human relations approach has, as we have said, become a part of the manager's everyday theory of organization.* In this form it has appeared alongside classical theory as another universal prescription for all managers and all organizations, in spite of the nonuniversalist stance of its principal founders.

Given the personal histories of the principals, and their time, it is not difficult to understand how they came to initiate the human relations approach and how it snowballed into a movement. The chief contributors, such as Mayo, Lewin, Roethlisberger, and McGregor, were all academics whose personal experience was largely with universities. Considering

* The almost exclusive attention given to communication processes in human relations as a popular movement should not obscure the fact that throughout this period a number of human relations researchers were calling attention to the influence of other important variables. For example, the impact of technology on organizational behavior has been a continuing research theme. This work, as carried out by such people as Charles R. Walker, William Foote Whyte, Eric Trist, A. K. Rice, and Leonard Sayles, has not had a widespread impact as yet on the human relations movement.

this exposure to a relatively unstructured type of organization, it is not surprising that they viewed the production segment of industrial organizations, where they began their systematic research, as very highly structured and restrictive. This is not to say that their findings were in any way invalid —only that their backgrounds would lead them to ask questions about the negative side effects of any highly structured system.

The trends of the times help to explain why these men's findings were grasped with such fervor and developed into a movement. The last three decades have, by common observation, been periods of increasing change and uncertainty, mostly stemming from the advances of science and related technology. Our own findings suggest that in such times the typical organization will need to reduce its degree of structure and lengthen its time horizon in order to remain viable. This has been especially true in the science-based industries, such as chemicals, electronics, and petroleum. It is no wonder that management has been turning to the human relations experts for guidance in reshaping their organizations to cope more effectively with change. In fact, the problems of change in organizations have been a persistent theme of human relations investigation. In studying the bank wiring room at Western Electric, the researchers found a combination of classical management practices plus a moderately high rate of technical change that disrupted relationships and kept alienating the work force from the purposes of the total organization. A good deal of the subsequent field research has been conducted in industries with high rates of change and particularly in engineering and research units, where change is especially rapid. It is in such settings that the findings of this study would indicate that a shift toward a wider distribution of influence would have a payoff in performance. Thus it is not surprising that the human relations researchers have found both support for their propositions and the most avid application of the implications of their findings in such organizations.[21]

In the light of these historical facts it is easy to see why some writers and many managers have picked up the human relations work as a universally applicable theory of organization. For example, in their eagerness to find a single answer to all industrial ills, some have interpreted the relay assembly test room experiment as offering evidence for using democratic procedures at all levels and in all parts of organizations. These people have overlooked the fact that in this original experiment participation did not go very far. Workers were consulted on such matters as the length and timing of rest periods and the seating arrangements. They were not brought into discussions of work scheduling, work methods, quality, payment, hours, or any other broader considerations. In fact, the workers were performing a short-cycle, highly repetitive task, which gave them a strictly limited base of knowledge in regard to organizational matters. And they actually achieved influence only in those limited areas where they had relevant knowledge.

Reinterpreted in the terms of this study, the relay assembly test room was an experiment at the worker level in giving people positional influence to match their knowledge-based influence, limited though it was. It worked out well. It enhanced the workers' identification with the goals of the department, gave them a sense of responsibility, motivated them —choose your own favorite terms—, and output went up and up. To call this experiment a universal argument for democratic (or nondemocratic) management, however, is simply confusing matters.

The same can be said of the White and Lippitt experiments, also widely cited as evidence for the universal utility of democratic leadership. Their careful study clearly demonstrated that democratic leadership provided a number of advantages over more autocratic or *laissez-faire* styles. However, as the authors point out, these were hobby groups designed to achieve the ends that the boys themselves wanted to achieve. They were recreational clubs. The boys were the source of all relevant information on what was desired. To design a structure and a leadership pattern that taps into this knowledge rather than repressing or ignoring it makes good sense, as the findings show. Broad generalization from this experiment is, however, to be done with great caution, since, as the authors stated, "The situation was also not comparable with the many situations in which society demands that a certain end be accomplished." [22]

It is possible now to restate the major tenets of the human relations movement (as distinct from its research) in the light of the present study. In our terms, the movement advocates the general use of a low-structure organization, along with widely shared influence and open, confronting modes of conflict resolution. It has placed almost all its emphasis on realizing a high state of integration and has definitely played down the utility and importance of concurrently achieving appropriate differentiation. Once we have stated the ideas of this movement in these terms, it is obvious that the findings of our study suggest conditions under which these propositions will and will not hold. We have seen strong indications that organizations with less formal structure and widely shared influence are best able to cope with uncertain and heterogeneous environmental conditions. All the organizations in our sample seemed to function best when influence was located (either vertically in the hierarchy or laterally between functions) where the relevant knowledge was concentrated. In some instances this meant that the lower echelons had considerable influence on organizational issues, and in some instances it did not. In all the organizations studied the confronting modes of conflict resolution seemed to contribute to effective integration. However, we did not study any circumstances when conflict was predominantly of a zero-sum or win-lose type. We will say more of this in the next chapter. So we see that the tenets of the human relations movement are, in terms of our findings, relevant to many but not all of the situations studied. In today's world of rapid change the prescription is undoubtedly right many more times than otherwise. But as a movement, it is still a universalistic remedy for organizations that seem to need differential treatment.

Our review of two major schools of organization theory has placed both in the perspective of the findings of this study. We have seen that each theory seems to apply in certain environmental conditions. In simplified terms, the classical theory tends to hold in more stable environments, while the human relations theory is more appropriate to dynamic situations. This realization helps to explain the historical paradox posed earlier in this chapter—the parallel persistence of these two theories over a period of at least three decades. Both were needed to explain behavior in organizations operating in distinctly different environments; one theory could not displace the other.

The debate is still going on. Today it is in terms of Theory X *versus* Theory Y, or mechanistic *versus* organic. Sociologists concerned with organizations discuss a similar dichotomy in terms of "rational" *versus* "natural" modes of organization. The former is traced to Weber, who emphasized the virtues of a bureaucratic structure for achieving organizational purpose and meeting other latent human needs.[23] Subsequent sociologists have pointed out the shortcomings of Weber's model and have emphasized the unplanned evolution of organizational forms and the multiplicity of organizational goals. This debate also continues.

These theoretical discussions are, of course, necessary and healthy among students of organizational life. Their spillover onto managers, however, has caused confusion that comes at a high price. Far too many administrators raised in one organizational setting and infused with the theory appropriate to it have wrought havoc by trying to apply it later in quite different settings. A small army of managers, trained and conditioned by classical theory, has tried at great cost to apply it in inappropriate settings. The opposite has undoubtedly happened too, but perhaps to a lesser extent.

Early in this chapter we saw a long list of management techniques currently in vogue, which we sorted into two rough sets. One set would tend to develop a more highly structured organization; the other, a lower one. One set would tend to increase the sharing of influence; the other, to concentrate it. Are managers choosing wisely the right time and place for the application of these techniques? Or are they being applied across the board, without discrimination? To repeat—the cost of confusion in these matters remains very high.

Our findings have led us to believe that both of the traditional theories might now be subsumed under a broader theory that is gradually taking form in the recent literature—what we are calling *a contingency theory of organizations.* We would hesitate to make such an assertion if we did not see a number of recent studies conducted by others that tend generally to support our conclusions. In the next chapter we will examine a selection of these and pull them together with our findings into a further statement of a contingency theory of organizations.

CHAPTER VIII

Toward a Contingency Theory of Organization

IN THE PREVIOUS CHAPTER we kept our review of organization theory within reasonable limits by restricting it primarily to the work of the early writers in both the classical and the human relations schools. Since those early days studies on human behavior in organizations have appeared in an ever-increasing tide. This flood of material is often the despair of practicing managers and students who try to keep up with it. Even skillfully written digests and reviews cannot adequately cover the available material. They usually end merely with a plea for a better theory that can synthesize the diverse findings.[1] This present chapter cannot fill this need, but we hope to make a contribution by examining a selected sample of important modern organizational studies in the light of our own research. We will attempt to find out whether a number of seemingly diverse studies can be fitted together into a sensible and consistent pattern that meshes with the present work. Can this pattern provide a needed direction for future research and contribute to an emerging new theory of organization?

The separate studies to be reviewed here were selected according to several criteria. First, and most obviously, we were looking for studies of organizations—of how organizations or major parts of them function, based on the systematic collection of empirical data. Second, we were interested in studies that, like our own, were multivariate. While researchers generally agree that modern, complex organizations need to be examined as multivariate systems, many current studies still address only two or perhaps three variables at a time. This is understandable given the limited research resources and access to data, but such studies are not especially useful for our present purposes. Third, we selected studies that are contingent in the sense that they try to understand and explain how organizations function under different conditions. The conditions of concern will naturally vary, but they should, for our purposes, come from "outside" the organization as usually defined—from the environment or context that the organization has chosen as its domain of operation. These outside contingencies can then be treated as both constraints and opportunities that influence the internal structure and processes. Finally, we wanted studies that differed widely in their approach. We were seeking diversity in research methods and in the underlying discipline and conceptual framework. Given the focus on organizations, it will come as no surprise that several of the studies reviewed here were done by sociologists, but there are also studies conducted by psychologists, social psychologists, a historian, and an economist. We will not attempt a complete picture of each of these studies, but will highlight those aspects that are relevant to our themes.

We will start with a look at six major studies, each of which makes a contribution to understanding what specific organizational attributes are related to certain external characteristics of the immediate environment or the nature of the organization's primary task. Later we shall examine other contingency studies—first with regard to conflict resolution and then some focusing on the predispositions of organization members.

CONTINGENCY STUDIES EMPHASIZING ENVIRONMENTAL AND TASK VARIABLES

In a broad way all of the six studies we will now examine attempt to throw more light on how organizations must vary if they are to cope effectively with different environmental circumstances. We will be interested in seeing how the results of these studies fit with our own findings. We will begin by looking at a significant piece of work done by Burns and Stalker, two industrial sociologists who had a major impact on the design of our study.

Burns and Stalker

Tom Burns and G. M. Stalker examined some 20 industrial firms in the United Kingdom.[2] For a detailed, in-depth study, this gave them an unusually good comparative sample. Their work focused on how the pattern of management practices in these companies was related to certain facets of their external environment. The particular external characteristics examined were the rates of change in the scientific techniques and markets of the selected industries. Next they explored the relationship between internal management practices and these external conditions to discover its effect on economic performance. The evidence was collected by extensive interviewing of key people in all 20 companies. No measurement methods were attempted. The enterprises were drawn from a variety of industries: a rayon manufacturer; a large engineering concern; a number of diverse Scottish firms, all interested in entering the electronics field; and eight English firms operating in different segments of the electronics industry. Fairly early in the fieldwork the authors were struck with the distinctly different sets of management methods and procedures they found in the different industries. These they came to classify as *"mechanistic"* or *"organic."* The following paragraphs summarize this aspect of the study:

There seemed to be two divergent systems of management practice. . . . One system, to which we gave the name "mechanistic," appeared to be appropriate to an enterprise operating under relatively stable conditions. The other, "organic," appeared to be required for conditions of change. In terms of "ideal types" their principal characteristics are briefly the following.

In mechanistic systems the problems and tasks facing the concern as a whole are broken down into specialisms. Each individual pursues his task as something distinct from the real tasks of the concern as a whole, as if it were the subject of a subcontract. "Somebody at the top" is responsible for seeing to its relevance. The technical methods, duties, and powers attached to each functional role are precisely defined. Interaction within management tends to be vertical, *i.e.,* between superior and subordinate. Operations and working behavior are governed by instructions and decisions issued by superiors. This command hierarchy is maintained by the implicit assumption that all knowledge about the situation of the firm and its tasks is, or should be, available only to the head of the firm. Management, often visualized as the complex hierarchy which is familiar in organization charts, operates a simple control system, with information flowing up through a succession of filters, and decisions and instructions flowing downwards through a succession of amplifiers.

Organic systems are adapted to unstable conditions, when problems and requirements for action arise which cannot be broken down and distributed among specialist roles within a clearly defined hierarchy. Individuals have to perform their special tasks in the light of their knowledge of the tasks of the firm as a whole. Jobs lose much of their formal definition in terms of methods, duties, and powers, which have to be redefined continually by interaction with others participating in a task. Interaction runs laterally as much as vertically. Communication between people of different ranks tends to resemble lateral consultation rather than vertical command. Omniscience can no longer be imputed to the head of the concern.[3]

Our findings, in a general way, strongly support these conclusions of Burns and Stalker. We systematically measured some of the structural attributes in which they observed variance, such as specificity of role description and reliance on formal rules. The environmental attribute that Burns and Stalker saw as causing the variance in management practice —the rate of change in technologies and markets—is one of the variables we have subjected to more systematic inquiry. We have found, as these findings suggest, that effective organizational units operating in stable parts of the environment are more highly structured, while those in more dynamic parts of the environment are less formal. The convergence of the findings of the two studies is considerable and important, even though Burns and Stalker's was an exploratory study on which we drew to develop a more complex research model.

Woodward

Another study that influenced our design was conducted by Joan Woodward, one of England's leading industrial sociologists, who in 1953 organized a research team and launched an investigation of the management process that, in various forms, is still continuing. This fruitful effort started with the question of whether "the principles of organization laid down by an expanding body of management theory correlate with business success when put into practice."[4] In order to address this broad question, the researchers chose a rather uncommon and bold strategy. They selected a geographical area (South Essex) and studied virtually all the firms (91%) in the area that employed at least 100 people. Thus they secured a sample of 100 firms in widely diverse lines of business and proceeded to make a rather detailed examination of their characteristics, with particular attention to management practices. Early in the study they began to realize that there was no significant direct association between the management practices of these firms and their business efficiency or their size. Accordingly, Woodward commented:

> The widely accepted assumption that there are principles of management valid for all types of production systems seemed very doubtful—a conclusion with wide implications for the teaching of the subject.[5]

The researchers then sought some other basis of accounting for the variations in management practices. They found that when the firms were grouped according to their techniques of production and the complexity of their production systems, the more successful companies in each of these groupings followed similar management practices. The three broad groupings were (1) small-batch and unit production, *e.g.*, special-purpose electronic equipment, custom-tailored clothing; (2) large-batch and mass production, *e.g.*, standard electronic components, standard gasoline engines; and (3) process or continuous production, *e.g.*, chemicals, oil refining. These three rough classifications were next broken down into 10 subgroups that, when arranged, formed a rough scale of the predictability of results and the corresponding degree of control over the production process. This ranged from low predictability for unit production to high for process production. As an example of predictability, the author points out:

> Targets can be set more easily in a chemical plant than in even the most up-to-date mass-production engineering shops, and the factors limiting production are known more definitely.[6]

The general conclusion of this study, as we indicated in Chapter I, was that the pattern of management varied according to these technical differences. This was especially true of the more successful firms. In other words, economic success was associated with using management practices that suited in specified ways the nature of the various techniques of production. To be more specific, Woodward found, for example, that the number of levels in the hierarchy and the ratio of managers to hourly personnel increased directly with the predictability of production techniques. In a less measurable way, this same predictability factor appeared to influence the type of coordination required between the basic business functions, the level at which decisions were made, and the relative dominance or influence given to the various functional units.

Again we can see a general convergence of Woodward's study and ours, even though her work centered on the organizational impact of variations in predictability only among production systems confronted with different technologies. Ours, on the other hand, emphasized the effect of variations in predictability on the tasks of all three of the basic units—production, marketing, and research—in industries that differed in the degree of market and scientific certainty, but that primarily used a process technology. Woodward's study, like ours, concludes, "There can be no one best way of organizing a business," and also provides strong leads as to how organizations must vary to be successful under different task and environmental conditions.[7]

Fouraker

With this third study of organizations we make an abrupt shift from the now-familiar approach of the sociologist to that of an experiment-oriented economist. We became aware of this work only after we had finished our own investigation. Fouraker began with an interest in the way organizations handle conflict and make economic decisions.[8] Starting from a very few basic premises about human choice behavior, he logically deduced how two polar organizational types would respond to different choice situations posed by their environments. For preliminary tests of the propositions thus derived, he and his colleagues devised some ingenious small-scale experiments that simulated the organizational types and the environmental circumstances. The experiments generally supported the theory.

Fouraker identified the two polar types as (1) the "L" organization, composed of highly independent management motivated by their own aspirations and members who are relatively responsive to the aspirations of others; and (2) the "T" type, composed of responsive management and independent members. He hypothesized that the pure types, made up of all independents or all responsive people, would be unstable and would therefore drift into either L or T types.

Fouraker describes the L organization as:

> . . . the classical or traditional form for organized human effort. The prospect (usually threatening) of a test with another group, or with nature, provides the common purpose for the original organization. The unity of interest implies that conflict within the group is dangerous and should be suppressed. This can be done, perhaps most easily, by selecting one person to act as the leader: his task is to select the appropriate goals or objectives for the group. . . .
>
> The L organization is authoritarian. It does not generate the social mechanism or management skill to tolerate or contain internal conflict. Discipline is a necessity to insure harmony of interest and outlook. . . .
>
> The L organization seems to be a very effective response to an institutional environment that is:
>
> 1. Fairly stable, or not complex;
> 2. Basically threatening.
>
> The requirement for simplicity or stability stems from the leader's role: the responsibility for choosing objectives is assigned to him, and no person can adjust to many simultaneous demands. The human decision maker tends to approach problems sequentially, not simultaneously.
>
> The threatening aspect is required to insure discipline within the organization. Discipline, as we have seen, is necessary to maintain conformity of values.[9]

The T organization, in marked contrast, consists:

> . . . of members who are independent technical specialists, with a responsive management. This organizational form is of relatively recent origin, since it depends upon the existence of:
>
> 1. Technical specialists in several dimensions,
> 2. Whose special output must be coordinated to achieve the objectives of the organization.
>
> Specialization emerges from, and causes, complex social situations. A complex task requires sustained attention, commitment, and interest. To master such a role one must learn, and ultimately become identified with, the task. Learning may lead to innovation and change. It is this contribution of the specialist which must be coordinated with the efforts of others in the T organization.
>
> To pursue his specialty effectively, the member must be independent. His identification is with his task and with those who pursue similar tasks. His loyalty to his discipline, and to others who pursue it, may be stronger than his loyalty to his organization. He has more in common with similar specialists outside his organization than he has with specialists in different dimensions within his organization. . . .[10]
>
> The management attempts to coordinate the efforts and output of these specialists. This requires a considerable exchange of information, and therefore numerous communications channels.
>
> A T organization is likely to have very little in the way of hierarchy or chain of command. The status distinction between management and members will not be great. Generally the management will be drawn from the membership, and occasionally may return. The symbolic activity in the T organization is the committee meeting. The committee is the major means for coordinating the efforts of the specialists. As developed before, the committee is the institutional structure designed to secure commitment from members to a common set of objectives. . . .[11]
>
> If the environment is favorable, the T organization is likely

to be extremely productive and effective. The independent specialists can identify and pursue opportunities:

1. Simultaneously;
2. At a rapid rate.

This contrasts with the limitations of the independent leader of an L organization confronted with a favorable environment.[12]

In describing the T organization Fouraker addresses the central issue of our research: the problem of conflict resolution in complex organizations that need to combine and integrate the work of highly differentiated departments to cope with heterogeneous and dynamic environments.

One apparent difference between our study and Fouraker's needs further examination. The characteristics of the immediate environment that we saw influencing the pattern of the organization were the rate of change and the predictability of events. Fouraker, however, focused on the "favorable or unfavorable" nature of the environment. These characteristics, on further analysis, are not so different as they might at first appear. To Fouraker a favorable environment was one in which new resources and new opportunities were becoming available for the organization to exploit. It would appear that this could only happen during a period of fairly rapid change. Likewise, his unfavorable environment is characterized by scarcity of resources, with a number of organizations struggling to secure their share. In such an environment, though the form of competition among organizations might change, the resources available, as he himself points out, are relatively static—hence, in our terms a stable, relatively unchanging environment. Given this rough reconciliation of the treatment of environmental characteristics in the two studies, we can see a general convergence of findings. In essence, both Fouraker's study and our own point to the different organizational characteristics required for effectiveness under different environmental conditions. This convergence is particularly interesting since Fouraker's findings were derived theoretically from a few basic premises and tested by some small-scale experiments, while ours were tested by empirical data from a sample of real organizations.

Chandler

In *Strategy and Structure: Chapters in the History of Industrial Enterprises,* Alfred Chandler, a historian, has created a landmark study of the evolution of large organizations.[13] His method is the comparative analysis of the case histories of a few pioneering firms, supplemented by a brief review of the administrative histories of nearly 100 other major American companies. The basic thesis is deceptively simple: that organization structure follows from, and is guided by, strategic decisions. He develops this theme by selecting for extensive examination Du Pont, General Motors, Standard Oil (New Jersey), and Sears Roebuck. He chose these four companies as the innovators in creating a successful structure for administering the large multidimensional enterprise. Since their organizational innovations have subsequently been widely imitated, as Chandler documents, these case histories broadened into the writing of institutional history. Chandler's four case histories present a wealth of detail on the processes, often trial and error, by which these companies created their complex structures. For our present purpose we will draw on only some of his major conclusions.

Chandler sees new strategic choices arising from environmental changes: "Strategic growth resulted from an awareness of the opportunities and needs—created by changing population, income, and technology—to employ existing or expanding resources more profitably."[14] He traces these strategic growth phases and their organizational ramifications in each of the companies and shows them as responses to changing environmental conditions. At one stage it was an "expansion of volume" that led to the development of more specialized jobs within the existing major functions. Then "growth through geographical dispersion" led to the establishment of territorial offices. Then the decision to become vertically integrated led to expansion into "new types of functions," with sharper differentiation of functional subsystems. Finally, diversification through the development of "new lines of products" led to the establishment of various product divisions. Chandler explores how each of these choices to create new organizational units led into problems of integration that in turn required the development of new integrative (or, in his terms, administrative) structures. In fact, Chandler argues, "whenever the executives responsible for the firm fail to create offices and structure necessary to bring together effectively the several administrative offices into a unified whole, they fail to carry out one of their basic economic roles." [15]

Throughout his study Chandler makes it clear that he sees different kinds of organization as necessary for coping effectively with different strategies and environments. Once again he cites the role of environmental change as the key factor in the choice of appropriate structure. One passage merits quoting at some length:

> When the managers of a federation or combination of integrated companies decided to coordinate, appraise, and plan systematically the work of their far-flung enterprise, they almost always consolidated these activities into a single, centralized, functionally departmentalized organization. As had the different mergers among the users of steel, such as International Harvester, Allis-Chalmers, American Locomotive, American Car and Foundry, and American Can, so many other combinations quickly disbanded their constituent firms and placed the sales, manufacturing, purchasing, or engineering activities of each within large, single-function departments. By 1909, only two or three of the nation's largest industrial empires remained pure holding companies. More, like Standard Oil, American Tobacco, and the meat packers,

continued to be legally both holding and operating companies that administered both functional departments and multifunctional units, the latter usually having the legal form of a subsidiary corporation. The large majority, however, became administered through centralized, functionally departmentalized structures. For until Alfred P. Sloan created a new organization form at General Motors in 1920, this structure appeared to be the only one which could assure effective administrative control over a large industrial consolidation.

Yet the dominant centralized structure had one basic weakness. A very few men were still entrusted with a great number of complex decisions. The executives in the central office were usually the president with one or two assistants, sometimes the chairman of the board, and the vice presidents who headed the several departments. The latter were often too busy with the administration of the particular function to devote much time to the affairs of the enterprise as a whole. Their training proved a still more serious defect. Because these administrators had spent most of their business careers within a single functional activity, they had little experience or interest in understanding the needs and problems of other departments or of the corporation as a whole. *As long as an enterprise belonged in an industry whose markets, sources of raw materials, and production processes remained relatively unchanged, few entrepreneurial decisions had to be reached. In that situation, such a weakness was not critical, but where technology, markets, and sources of supply were changing rapidly, the defects of such a structure became more obvious* (italics added).[16]

Unlike the authors discussed previously, Chandler focused on the big and relatively infrequent strategic shifts in major corporations. In this sense he was not interested in the differences created by technologies, functional specialization, or environmental congeniality. Nevertheless, he concluded that different environmental conditions demanded different structures. To Chandler also, it was the rate of environmental change (in "technology, markets, and source of supply") that created the pressure for strategic and subsequently structural change. This historical study adds a new and different source of support to our enlarging picture of a theory that recognizes the contingent relationship of organizational characteristics to environmental demands.

Udy

Stanley Udy, a sociologist, employed a strikingly different way of examining the relationship between technology and organization structure. His purpose was "to seek broad generalizations about variation in organization structure relative to its social setting and the technology involved; and thus contribute to filling some of the 'gaps' in organization theory." [17] He decided to study nonindustrial societies and drew his evidence primarily from the Human Relations Area Files, a compilation of anthropological descriptions of some 150 separate societies. From this source Udy developed a sample of 426 organizations that were carrying out various forms of agricultural work, hunting, fishing, collection, construction, manufacturing, and stock-raising. The societies represented all parts of the world, all major social groups, and several widely separated periods of history. He categorized the attributes of each of these organizations as well as the technology and the social setting.

Udy's major conclusion had to do with the strength of the association between organization and technology:

> Given a systematization of the possible range of variation of technological processes, it was found that certain aspects of authority, division of labor, solidarity, proprietorship, and recruitment structure could be predicted as to general trend from technology alone.[18]

Across the full sweep of the known nonindustrial societies, Udy's evidence clearly indicates that the facts of technology alone have a distinct and persistent influence on the structure of viable organizations.

Since Udy focused on organizations doing nonindustrial tasks, probably under relatively stable technical and market conditions, we cannot make direct and specific connections between his study and the others described here. His very broad-based work does, however, lend impressive support to the very general conclusion that organizations doing different tasks must be structured differently. Beyond this general point, his findings are particularly relevant for the design of any modern international organization operating in many cultural settings—the implication being that the technical requirements of the task cannot safely be ignored in designing the organization structure, even as allowance is made for cultural differences.

Leavitt

Shortly after World War II, at MIT, Alex Bavelas began another set of research studies that, in different forms, is still continuing. Although many others have contributed to the work, we will draw on the writing of Harold Leavitt, a social psychologist, who not only conducted some of the key experiments but has gone further than the others in considering the study's implications for management practice.

Leavitt and his colleagues used small groups to conduct various problem-solving activities under experimentally controlled conditions. The situation with which they experimented most extensively involved five people, each of whom was given a cup containing five different-colored marbles. In each cup was one marble that was duplicated in all five cups. The people were asked to exchange written communications until all five had learned which color marble they had in common. The experimental variations on this problem were introduced by controlling the channels available for communication (see Figure VIII–1). Using one of these three net-

Figure VIII–1

THREE COMMUNICATIONS NETWORKS

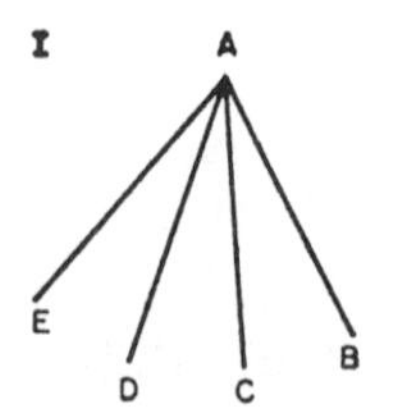

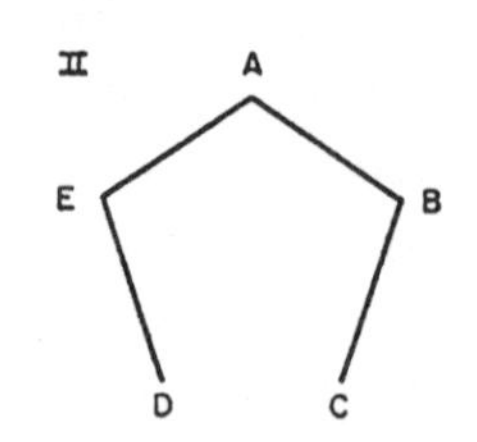

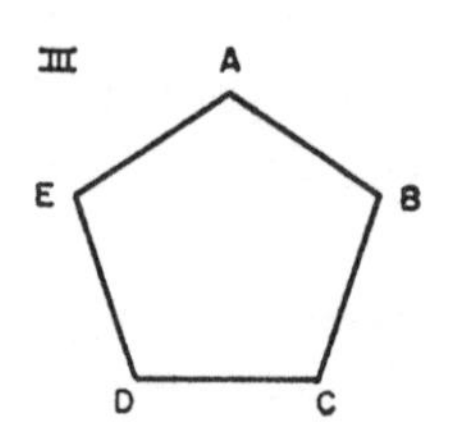

works, each group worked through the problem again and again, with a new set of marbles each time. Records were kept on a variety of results—the speed of reaching solutions; the number of messages sent; the number of errors made; etc. Without going further into the details of these experiments we can examine the findings. Leavitt has summarized them as follows:

> It turns out that on these simple tasks Network I is far more efficient than II, which in turn is more efficient than III. In other words, groups of individuals placed in Network I within a very few trials solve these problems in an orderly, neat, quick, clear, well-structured way—with minimum messages. In Network III, comparable groups solve the same problems less quickly, less neatly, with less order, and with less clarity about individual jobs and the organizational structure—and they take more paper too.[19]

However, when the researchers asked their subjects how they felt about their experiences, they got quite a different picture:

> Network III people are happier, on the average, than II or I people (though the *center* man in Network I is apt to be quite happy).

Furthermore, the experimenters noticed:

> . . . when a bright new idea for improvement of operations is introduced into each of these nets, the rapid accep-

> tance of the new idea is more likely in III than in I. If a member of I comes up with the idea and passes it along, it is likely to be discarded (by the man in the middle) on the ground that he is too busy, or the idea is too hard to implement, or "We are doing O.K. already; don't foul up the works by trying to change everything!"

These observations led to an additional experimental change. The researchers introduced "noisy" marbles—marbles of unusual colors for which there were no common names. They again found that Network III had certain advantages over I:

> . . . Network III is able to adapt to this change by developing a new code, some agreed-on set of names for the colors. Network I seems to have much greater difficulty in adapting to this more abstract and novel job.

Leavitt summarized these findings:

> So by certain industrial engineering-type criteria (speed, clarity of organization and job descriptions, parsimonious use of paper, and so on), the highly routinized, noninvolving, centralized Network I seems to work best. *But* if our criteria of effectiveness are more ephemeral, more general (like acceptance of creativity, flexibility in dealing with novel problems, generally high morale, and loyalty), then the more egalitarian or decentralized Network III seems to work better.

These experimental findings fit very neatly with those of the present study and the others we have reviewed. It takes different kinds of organizations to perform different kinds of tasks efficiently. Of all the researchers we have reviewed, Leavitt sees the practical implications of these findings most clearly. He states:

> . . . these developments suggest that we need to become more analytical about organizations. . . . and to allow for the possibility of differentiating several kinds of structures and managerial practices within them.

* * * * *

> . . . I suggest that more and more we are differentiating classes and subclasses of tasks within organizations, so that questions about how much we use people, the kinds of people we use, and the kinds of rules within which we ask them to operate, all are being increasingly differentiated—largely in accordance with our ability to specify and program the tasks which need to be performed and in accordance with the kind of tools available to us.
>
> * * * * *
>
> . . . such a view leads us toward a management-by-task kind of outlook—with the recognition that many subparts of the organization may perform many different kinds of tasks, and therefore may call for many different kinds of managerial practices.

Leavitt also anticipated that the more differentiated organizations he sees emerging will "make for more problems of communication among sets of subgroups." But his experiments did not permit him to explore these integration issues further. For our present purposes, however, the impressive point about this entire line of research is that studies on an ostensibly different topic, *i.e.,* problem solving, and using a completely different methodology, *i.e.,* small-group experimentation, should come up with such parallel findings about the relation between organizational attributes and the characteristics of different tasks. Leavitt's phrase "management-by-task" serves well to capture the implications of this research.

The six studies we have reviewed have all been concerned with the various ways in which organizations, or major parts of organizations, are designed in terms of structure and important management practices, and the contingent relation this bears to their performance of different tasks in different environments. Of particular importance are some of the apparent differences our review highlights in the way the "external" conditions have been conceptualized and made operational in each study. While there are differences in terminology, our review suggests that these are referring to the same underlying phenomena. Further conceptual and empirical work is needed, primarily on these external variables as a further test of our way of reconciling these studies. In any event, the studies, taken together, offer a formidable body of evidence that different organizational forms are required to cope effectively with different task and environmental conditions. All represent a strong new trend toward contingency and comparative studies of organizations that can serve to reconcile both the classical and the human relations approaches.

OTHER TYPES OF CONTINGENCY STUDIES

Conflict Resolution

Another major theme of our research is the process of conflict resolution and the relation between states of differentiation and states of integration. Is there a comparable body of related research in this area? The answer is yes and no. Yes, there is a rich and growing literature on the dynamics of different kinds of conflict resolution. We will make no attempt here to review this literature except as it helps us to broaden our own view of the subject. The answer is mostly no, however, when it comes to studies of conflict resolution under various conditions of organizational differentiation. Some of the studies we have just reviewed provide the few limited exceptions. We shall briefly consider their relevant points, along with the prospects for future research on this subject.

In starting this discussion, we shall indicate that the general literature on conflict resolution has found it useful to distinguish between two general classes of conflict. This distinction is especially important for gaining perspective on the present research, since all the conflict we have considered has fallen into only one category.[20]

The first and most widely researched class of conflict arises when each of two parties has an interest in an issue such that any gain for either is at the expense of the other. This situation has been called a zero-sum competitive game, a pure-conflict game, and an issue. The theoretically most effective strategy for dealing with this situation is the use of bargaining, in which each party starts with a public position favorable to itself, and each alternately makes concessions in exploring for any overlapping ground where agreement can be reached. This form of bargaining—called "distributive bargaining" by Walton and McKersie [21]—involves withholding information about one's own goals and aspirations while trying to elicit this same information from the other party. Wage bargaining between unions and management is the classical example of this type of conflict.

The other class of conflict involves a problem to which many solutions are possible. The potential benefits for the parties involved are not fixed. Though interests may conflict, there is always the possibility of finding an ingenious solution that minimizes these conflicts. The disagreements that arise out of selecting a new product idea from an open-ended set of alternatives is a perfect example of this type. Many, many items of information may bear on such a problem. The various parties may have favored alternatives, but the theoretically best strategy for resolving the conflict involves a rapid and complete sharing of available information, including the weights given to different selection criteria. Such an exchange would be followed by a joint search through the shared information for alternatives that best satisfy the combined selection criteria. Implicit in this process is agreement among the parties on basic goals at some higher level of abstraction, if not on the more immediate means to these ends. Walton and McKersie have termed this process "integrative bargaining." [22]

All the conflict situations we have considered in this research have been of this second type. The partial determinants of successful conflict resolution that we have empirically tested have a demonstrated relevance to this class of conflict. The usefulness of openness and confrontation is probably severely limited in zero-sum conflicts. The broader research on conflict resolution has thus given us a necessary conditional modifier to our findings.

As we have said, there has been relatively little research besides ours on how differences of organizational form influence the conflict-resolution process. The present study has clearly indicated that the degree of intramural difference in structure and orientation significantly affects integration and the resolution of interdepartmental conflict. We have seen that these differences are crucial in situations where there is no conflict of basic goals. Certainly they can be expected to further compound the resolution process where goals do conflict. Both the Woodward study and the Burns and Stalker research explored this topic, but only in a general descriptive way. Of all the studies already cited, Fouraker deals most systematically with the conflict-resolution process. Using his L (authoritative) and T (technical specialists) typology, he deductively demonstrates how the L type is suited to carry out distributive bargaining successfully, but how it would be at a severe disadvantage in an integrative bargaining situation. The T organization would reverse this pattern of advantages and disadvantages. He also hypothesizes about what happens when an L organization confronts a T organization in regard to each class of conflict situation. It is this type of proposition that needs further testing. A new and challenging line of research that takes account of organizational differences as contingent variables is now possible on conflict resolution. It would be especially pertinent for the needs of many modern organizations to conduct such studies on conflicts that arise among elements of multidivisional and multinational companies.

Individual Predispositions

Our own study and the others we have reviewed so far have focused on the contingent relationship between the internal characteristics of the organization and the demands of its external environment or its task. There is, however, another important variable, which we have treated as a minor theme. This is the relationship between the kinds of predispositions members bring to their jobs and the consequences that various organizational characteristics have on these members as individuals. Of the many modern studies of organizations that center on individual attributes, three especially highlight the contingency idea and show the futility of searching for universal answers.

Fiedler. We spoke earlier of Fiedler [23] as the developer of the instrument for measuring interpersonal style, which we have used in this study. Fiedler himself has made extensive use of this instrument to study the relations between leadership style and group performance under a variety of conditions. He varied conditions in three ways: (1) the simplicity or complexity of the task; (2) the prior feelings of like or dislike between the leader and his group; and (3) the amount of traditional power at the disposal of the leader. He studied the interplay among these variables in a considerable number of live and experimental groups. Without going into specifics, his general finding was that different kinds of leadership style paid off in high group performance under different conditions. The task-oriented style was associated with high performance under the extreme conditions, that is, when situations scored either very high or very low on his three contingent variables. The relationship-oriented style paid off for the middle-range conditions. This research clearly implies that there is no universally useful leadership style. Managers come into organizations with predispositions for using different interpersonal styles, but each style can contribute to performance under certain task and organizational circumstances. Perhaps eventually the results of Fiedler's study can be linked with ours in applying the implications of both to the practical issues of organization design and staffing.

Vroom. In studying the reaction of different workers in a large parcel-delivery organization to the use of participative management methods, Vroom developed a similar contingency idea. In summary, he found that these employees had certain personality traits that predisposed them to respond (in terms of both attitude and performance) positively or neutrally to participative management. Vroom concludes:

> The present study corroborated previous findings that participation in decision making has positive effects on attitudes and motivation. It was demonstrated further that the magnitude of these effects is a function of certain personality characteristics of the participants. Authoritarians and persons with weak independence needs are apparently unaffected by opportunity to participate in making decisions. On the other hand, egalitarians and those who have strong independence needs develop more positive attitudes toward their job and greater motivation for effective performance through participation.[24]

Vroom's contingent variable, personality predispositions, is quite different from the task-environmental variables we have seen before. But the basic idea of studying the conditions under which different organizational practices are useful is the same, and is in sharp contrast to any search for universally valid practices. We should also note that Vroom's findings are consistent with our less systematic observations on the differences in predispositions of people in our high-performing plastics and container organizations.

Turner and Lawrence. A finding somewhat similar to Vroom's was reported from a study Turner and Lawrence did of workers' response to different job designs.[25] From a survey of 50 jobs in 11 industries they found that there was no universal response to variations in job complexity, which was determined by the built-in variety, autonomy, responsibility, knowledge, required interactions, and optional interactions. Rather, they found that two sets of workers, whom they labeled "city" and "town," brought into the factory persistently different orientations about the meaning of work. One group saw it as an unpleasant means to a desirable end. To them work chiefly meant putting in time and effort for an economic reward. The other group tended to see work as an end in itself, as an opportunity to express oneself through exercising a complex skill. It was only when these predispositions were considered that the researchers could account for the relation they found between job complexity and workers' satisfaction with their jobs. Once again we see the relation between an important organizational attribute, the complexity of industrial jobs, and employee response as contingent on some of the basic predispositions of workers.

We offer these three studies as examples of research that treats human predispositions as contingent variables in the study of organizational characteristics and outcomes. While other researchers, in a similar vein, are studying latent motivation patterns and the influence of aging and education, more research along these lines is certainly needed to generate a more comprehensive contingency theory of organizations. Such a theory will not only relate environmental characteristics to organizational attributes, but also will connect these variables to the varying predispositions of organization members.

A CONTINGENCY THEORY

We have sought to move beyond the picture presented in the preceding chapter of the strengths and limitations of the kind of organization theory most managers are still using—a disjointed combination of classical and human relations theories. By examining and relating some selected modern studies, we have tried to illuminate the prospects of a new research-based approach that we have tentatively labeled "contingency organization theory." This review has given support and further complexity to our own model, which starts with the examination of the interplay between any major part of an organization and its relevant external environment. The environment with which a major department engages is decided by the key strategic choice, "What business are we in?" Once that decision is made, whether explicitly or implicitly, the attributes of the chosen environment can be analyzed. To the environmental variables that we have examined the other studies cited have added the obviously important variables of the collective predispositions of the people (managers, specialists, hourly employees, etc.) who are drawn into the system from the environment. Internal attributes of the organization, in terms of structure and orientation, can be tested for goodness of fit with the various environmental variables and the predispositions of members. Unit performance (which will have to be judged by a number of dimensions, of which profitability is only one) emerges as a function of this fit.

This starting model is complicated as soon as we move to a complex, multi-unit organization in which each unit strives to cope with a different part of the environment. As soon as this happens, it introduces the complication of integrating the work of the different units. We see the existence of an integrating unit and conflict-resolution practices as contributing to the quality of integration and, in turn, to the overall performance. The resulting model does simplify the complex realities of modern organizations, as any model should, but it lends itself to further elaboration as needed. Each variable has been conceptualized and operationalized, at least crudely, for measurement. From our study and those discussed above we have derived some knowledge about the relationship among these variables. More research can and must be done on all aspects of the model. Even now, it does reflect the findings of a number of modern organizational studies. It clearly points the way to a more sophisticated model that will not only reduce the confusion in organization theory, but will also have considerable implications for the design and management of complex organizations. If such a model, as it evolves and improves, comes into any general usage, it will have impact beyond the boundaries of the specific organizations where it is applied. These practical implications of the contingency approach to organizational study will be examined in the final chapter.

CHAPTER IX

Implications for Practical Affairs

IN THIS FINAL CHAPTER we shall explore some practical applications of the ideas and findings we have reported. We shall focus on the implications for the design of effective organizations, and the implications for viable organizations in the future. We shall not dwell exclusively on our own research, but will draw at times on the broader thrust summarized in the discussion of the contingency study of organization.

As we move beyond factual findings to examine these practical implications, we must recognize two hazards. Stating implications of research findings always involves making a deductive leap of the order of saying that if A–B–C are true, it must follow that A_1–B_1–C_1 are true. There are bound to be uncertainty and chance of error in taking such leaps: Have variables D, E been accounted for? Will the relationship hold under other sets of conditions? etc. In the light of such uncertainty some would argue for a division of labor, with one set of people developing solid research knowledge while another group works on its implications. Otherwise, how can the high standards of scientific objectivity be maintained? The danger of separating research and application is that each group can easily get into a collective monologue, with no dialogue between them. When this happens, we can easily fail to put research results to use. Though there are legitimate grounds for differences of judgment on this matter, the authors have a long-standing concern with both theory and practice and have therefore chosen to address both subjects. The reader can, by being aware of the risk, minimize its dangers by doing his own double-checking of the deduction process.

The second hazard in discussing implications is that the act of applying knowledge to real-life affairs involves not only deductive analysis but also value choices. The expression "the designing of effective organizations" carries the implicit assumption that this is a valued objective. Some today would seriously question this. For the authors it is important, but by no means a monolithic value. Once again, though, it seems to us that, without getting deeply into a debate on values, an explicit awareness of the issue goes a long way in the present context toward handling the old charge that social scientists are the "servants of power." We are making no prior assumptions about the reader's own value system.

Implications for the Design of Effective Organizations

A groundswell of opinion is developing these days among some applied behavioral scientists and managers that the time is ripe for an infusion of more systematic and scientific techniques into the intuitive art the experienced manager practices as he designs and plans complex organizations. This argument has been well developed by Bennis [1] and is explicit in the provocative title of a recent book by Perlmutter, *The Social Architecture of Essential Organizations.*[2] This is, of course, the most general and pervasive implication of the research reported in this book. The entire thrust of the research is toward the more intelligent tailoring of organizations to their task and environments. Putting ourselves in this designing-architectural mood, we will now attempt to envision some specific ways of using our findings. For example, how can we more effectively subdivide the organization's overall tasks? What methods for achieving integration should we prescribe for a particular organization? What reward and control systems will motivate managers to achieve both the differentiated and the integrated performance required?

Applying the Differentiation and Integration Approach

As organizations undertake more complex tasks, they tend to complicate internally by differentiating new organization units. This process can readily be traced historically in many organizations. Spencer long ago noted this phenomenon in his analogy between social and biological systems:

> A social organism is like an individual organism in these essential traits; that it grows; that while growing it becomes more complex; that while becoming more complex, its parts require increasing mutual interdependence; that its life is immense in length computed with the lives of its component units . . . that in both cases there is increasing integration accompanied by increasing heterogeneity.[3]

The manager who develops a working familiarity with the concept of organizational differentiation will be more observant of this process. He can use the concept to sharpen his ability to discriminate intelligently between differentiation proposals based on new task complexities and superficially similar proposals based on Parkinsonian tendencies toward empire building. Our research indicates that the clear-cut and formal differentiation of organizational units, when based on significant task and environmental differences, contributes to good performance. By contrast, one common organizational error is to combine two distinctly different tasks in a single organizational unit. These conditions are schematically presented below (Figures IX–1 and IX–2); the solid lines depict the configuration of the actual formal structure and managerial orientation, and the broken lines represent the requisites stemming from the nature of the task and environment.

We found some evidence of this type of differentiation error in the low-performing food company. This firm maintained research as a single structural entity, in spite of indica-

Figure IX–1

APPROPRIATE DIFFERENTIATION

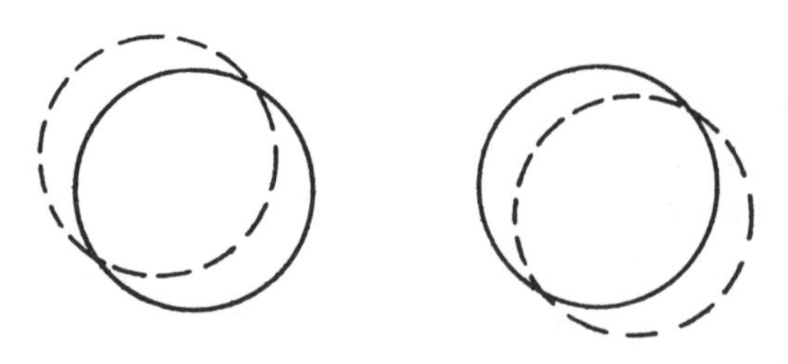

Figure IX–2

TYPE 1 DIFFERENTIATION ERROR

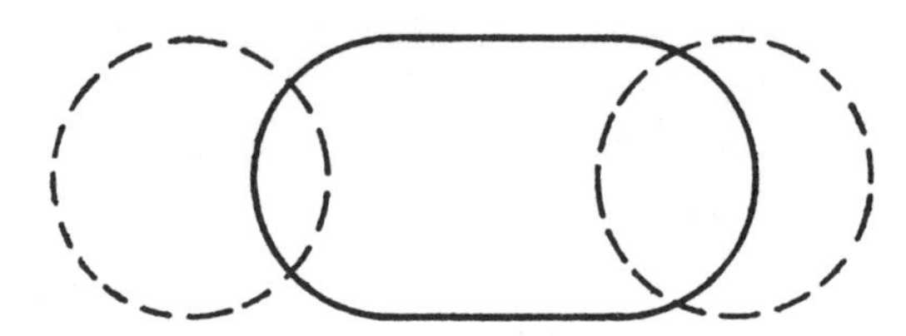

tions that the applied and basic research tasks required capacities to deal with somewhat different parts of the environment.

A historical example close to the experience of many businessmen may clarify this point. For years many companies treated selling as one big, unified task and the sales department as a management superstructure built up over the individual salesman. Most modern firms now recognize that the term sales covers over some distinctly different tasks—such as product development, product management, advertising and promotion, and market research, as well as sales management. The organizations that recognize these emerging task differences at a proper time and reflect them in their structure and related management practices tend to achieve a competitive advantage. But they must first break through the semantic barrier of seeing sales as a single entity. We think an understanding of the concept of differentiation developed in this study can help managers to solve similar organizational issues in the future.

The same process has led firms to break out the function of accounting into such different tasks as financial accounting and statistical control. Nowadays, operations research and computer specialists tend to be lumped together in one organizational unit, but an analysis of the tasks as they emerge might turn up the requirement for some considerably different orientations and management practices for each group.

The analysis of tasks and environments can also occasion a reversal of the splitting process. It sometimes reveals the wisdom of melding two units that have traditionally been separate. This has happened in some engineering departments, when task distinctions that were important at one period of time disappear with new knowledge. Our findings in low-performing plastics organization B indicated that their two separate applied research units were charged with essentially the same task. They had a record of competitive clashes, redundancy of effort, and poor coordination. This type of differentiation error is illustrated in Figure IX–3.

Figure IX–3

Type 2 Differentiation Error

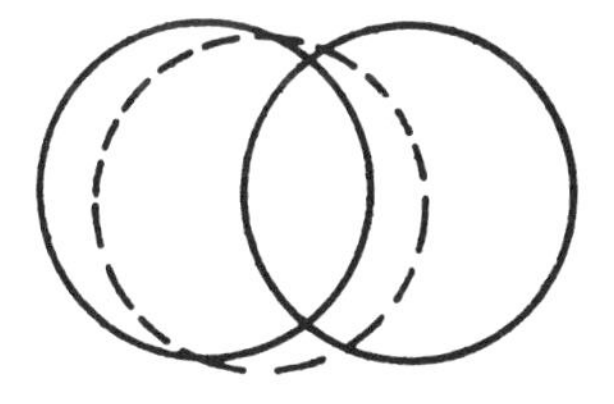

Any attempt to use the decentralization and integration approach systematically would have to begin with a diagnostic study of the organization and its immediate environment. This may seem an obvious point, but it is surprising how many organization planners start elsewhere. The first stage of such a study would involve examining the essential nature of the selected tasks and parts of the environments. What are the rates of change, the certainty of information, the time span of feedback, the relative number of environmental opportunities, the complexity of the tasks? These data can be gathered both from direct evidence and from the informed judgment of experts. Such information would provide a picture of the requisites for the differentiated units of the organization. Does it set up a requirement for heterogeneous or homogeneous units? Does it suggest some new splits or mergers among units? The study would then proceed to examine the actual attributes of the various units, along their crucial dimensions (which should not necessarily be the same as those used in this study). What are the structural characteristics? What are the actual orientations of the various units' members? Which units have structures or orientations that are out of line? Is it possible to pinpoint these problem spots? Our present diagnostic tools for these purposes are still crude but can be refined with further usage.

So far, in considering the application of the differentiation and integration approach, we have encouraged readers to take a designer's attitude toward organizations. This could also be called the social-engineering mood. It is not enough by itself. When it comes to application, the orientation of the teacher is also essential. One potential value of the differentiation and integration outlook lies in its use as an educational tool. When people live day in and day out in a specialized role, they tend to see their own organizational surroundings in terms of that role. The more personally involved in their jobs they become, the more this is true. Such involvement often leads them to personalize the conflicts that arise with representatives of other organizational units. Of course they know logically that an organization needs different kinds of specialists, but they forget the full meaning of this when they run into a particular person who is "impossible to work with." Then they all too readily turn to an explanation based on personality traits that writes off the individual as an oddball and justifies their own withdrawal from or forcing of the conflict.

Training sessions that build on differentiation and integration concepts and findings can be developed to improve this situation. Specialists can be trained in some depth to appreciate the fuller implications of the behavior of other types of specialists. The purpose underlying such training would be to help the participants to understand the reasons behind differences in orientations and behavior patterns and thereby to legitimatize and maintain them. Such training cannot erase the basic antagonism between differentiation and integration, but it can relax the tension somewhat.

As various specialists sit together and exchange firsthand information about their respective ways of working, the insights begin to appear. The mood of such meetings is not unlike that of a sensitivity training session, as people learn more about one another and the reasons for their differences. These exchanges can take place within a framework that is clearly related to task accomplishment. In such a setting applied researchers discover with a shock how differently they often think and work from their "fellow researchers" who are involved in more basic projects. The marketing managers are surprised to learn that an offhand response to a question from a basic researcher about the market potential of an idea might start the latter off on a six-month inquiry. If such personal exchanges can be tied to the quantified results of a general study of the state of differentiation and integration in the participants' own organization, the effect is reinforced. Thus the findings of a diagnostic study can be used for both planning and education.

Any study of the requisite and actual states of differentiation in an organization would be inadequate without simultaneous examination of the state of integration. The inverse relation this research has shown between the two states suggests the importance of any manager's thinking out the integration consequences of his differentiation plans. The whole trick is to plan concurrently for both conditions. Where it is necessary to develop highly differentiated units to cope with a diverse environment, provision must also be made for appropriate conflict resolution processes, or integration will be inadequate. Measuring the actual state of integration between each pair of units indicates the trouble spots where particular attention should be given to the development of integrative devices and conflict resolution procedures. This topic will be explored in the next section.

Applying Conflict-Resolution Findings

As we pointed out earlier, one traditional way organizations have resolved conflicts is by referring them up the line to the first superior shared by both disputants. This is the way the classical theorists assumed that all integration would be achieved. As organizations have undertaken more complex tasks in more heterogeneous environments, this procedure has not been adequate. Yet the findings of this study clearly show that it is still not only adequate but apparently highly effective under environmental conditions like those we found in the container industry. This condition is schematically shown in Figure IX–4 to indicate that modest differences in structure and orientation can be linked by a common boss.

One obvious type of error in planning the process of conflict resolution arises when such coordination through the management hierarchy is used as the only means of integrating units that are highly differentiated. This error is diagrammed in Figure IX–5.

It is interesting that we found no example of such an error in the 10 organizations we studied. Apparently managers have learned from experience not to try direct bridging of a wide gap. In fact, the error we found in the low-performing container organization was, in effect, the reverse. Here an inte-

Figure IX–4

EFFECTIVE CONFLICT RESOLUTION

(Direct linkage for mildly differentiated units)

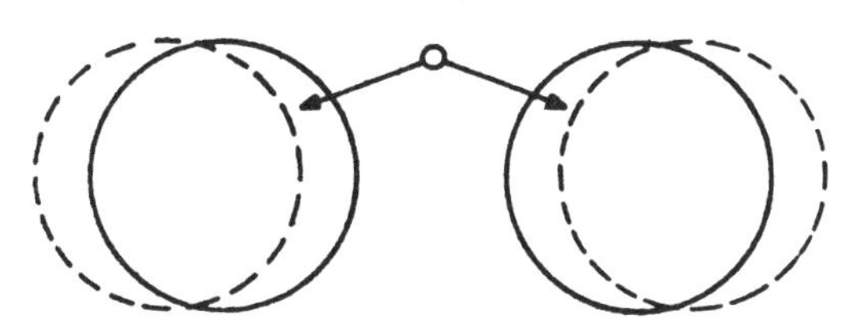

Figure IX–5

TYPE 1 CONFLICT RESOLUTION ERROR

(Ineffective direct linkage of highly differentiated groups)

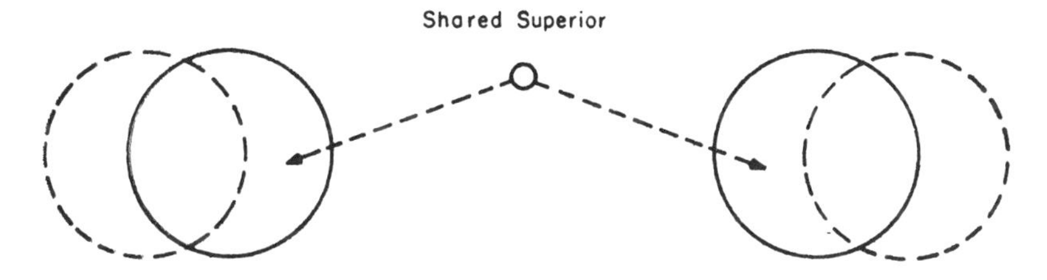

grating structure was set up to help settle disputes between units that were not sufficiently differentiated to justify it. Such groups seemed to add noise to the system and to make integration more difficult. This kind of error can be diagrammed as shown in Figure IX–6.

The absence of Type 1 errors and the presence of this Type 2 error could be interpreted as the result of a current management fad for decentralization, since establishing an integrating unit tends to push influence further down the structure.

The final error that arises from the use of integrating structures was also observable in the companies we studied—especially in low-performing plastics organization A. It resulted

Figure IX–6

TYPE 2 CONFLICT RESOLUTION ERROR

(Ineffective linkage from use of integrating unit between mildly differentiated units)

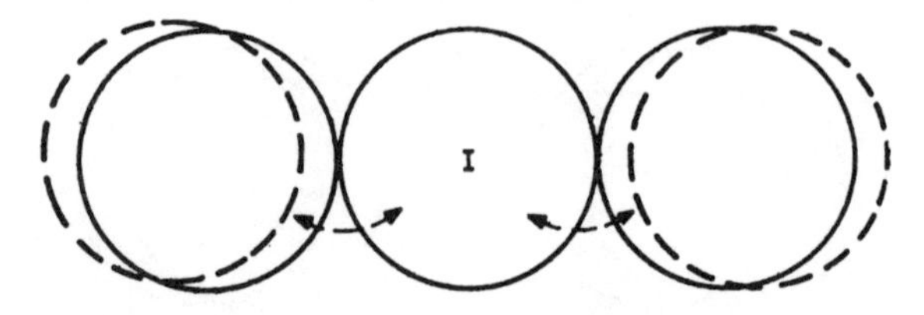

from an integrating unit being structured and oriented so nearly like one of the basic units that it lost its contact with the other. Figure IX–7 compares this type of error with the effective use of an integrating structure.

While this review of the possible types of errors in conflict

Figure IX–7

TYPE 3 CONFLICT RESOLUTION ERROR

(Ineffective linkage from misplacement of integrating unit)

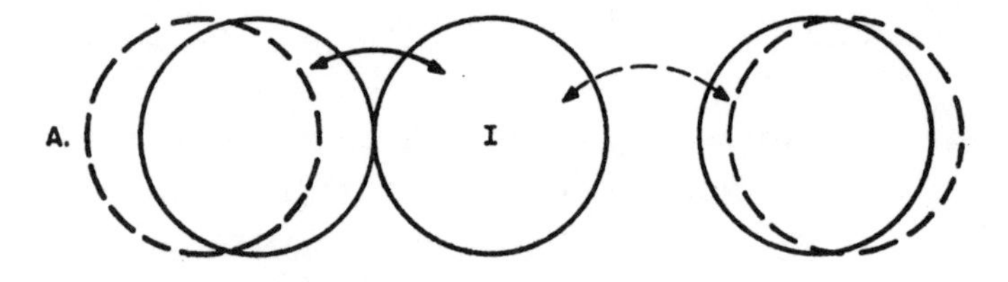

EFFECTIVE CONFLICT RESOLUTION

(Indirect linkage through integrating unit for highly differentiated units)

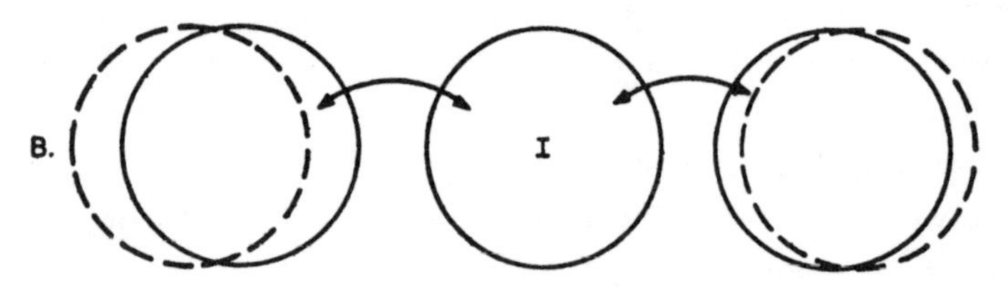

resolution is suggestive, it does not adequately reflect the range of experimentation that organizations are engaged in to solve their integration problems. We have seen in our data several variations on the use of cross-functional product teams and liaison roles. Industry is teeming with such efforts. One very interesting experiment, used primarily in the aerospace industry, is the matrix organization. In this form, which Gulick actually suggested over 30 years ago, each member of the organization is assigned two bosses. One is the functional boss of his specialty, while the other is the boss of his current work project. Neither is given arbitrary power to resolve conflicts. Thus each member of the organization is associated with two work groups, one made up of fellow specialists, and the other, of his co-workers in the diverse specialties required for a single project. Neither group can claim all of the man's attention or loyalty. Such an organizational form is an attempt to develop men with the appropriate specialists' orientation plus an involved enthusiasm for integrated project objectives.

The many conflicts in such organizations are expected to be handled among the members themselves. Each man has to make constant choices between his interests as a project team member and as a specialist. The matrix approach, while often useful, has distinct limitations, since it can at best link together no more than two types of differentiation. The usual matrix linkage of functional with product differentiation does nothing about the further complexity of differentiation based on such factors as physical space and customer types. To understand the possible problems we need only imagine a four-, five-, or six-dimension matrix organization. All these experiments, however, represent a lively interest in searching for new and better ways to resolve conflict and make decisions.

The more systematic way of understanding conflict resolution suggested by our research offers no pat solutions to these considerable complexities. It does, however, help to open the way toward more deliberate planning of how integration is achieved. Methods similar to ours can be used not only to help locate the points of chronic difficulty but also to help in identifying some of the important immediate determinants that are subject to management influence. Have the requisite interdependencies been mapped out? Is the influence located where the requisite knowledge is available—both vertically and horizontally in the system? Are differences confronted in an open, problem-solving manner, or is forcing or smoothing more predominant? Are integrating units used where needed (and not used where they are redundant) ? Are such units appropriately intermediate in structure and orientation? Are the integrators' rewards tied to the larger goals of the system? Do the integrators base their influence on expertise more than on formal power? These, of course, are the key questions that this research indicates could usefully guide a more systematic approach to the design and improvement of the conflict resolution process.

Once again, educational approaches can play an important part in applying the research findings. Managers in all the organizations we studied almost unanimously saw confrontation as the most *desirable* mode of conflict resolution. Yet our findings indicate it is used much less than it is recommended. This is most commonly explained by the assumption that people have the requisite knowledge, but have a personality-based aversion to confronting differences sharply. Our study offers a reminder that people may also not confront conflict because they do *not* have the requisite knowledge and yet feel a need to be influential. The two explanations would clearly call for different educational methods. There are management-development programs specifically designed to be helpful in the former situation.[4] The latter could be identified by a thorough diagnosis and could be addressed either by feeding the necessary knowledge to the person concerned or by changing the organizational expectations about who ought to influence particular kinds of decisions.

In either case it is important to recognize that one concomitant of confrontation is interpersonal competence. Determining to what extent persons can be trained to be more effective interpersonally is complex. Without getting involved in a lengthy discussion, we might note that one school of thought argues that skill in human relations is a central aspect of the individual's personality and cannot easily be altered.[5] Other behavioral scientists take the view that interpersonal competence can be developed through sensitivity training and other laboratory methods.[6] After studying the limited number of available systematic assessments of such training methods, we have come to recognize that the training methods can improve interpersonal competence and the ability to confront conflict to some extent. These experiences do not alter the managers' underlying personality characteristics, but they can alter their expectations of themselves and others about what is legitimate behavior, to the point that they are encouraged to behave more openly and to resolve conflicts more effectively.[7]

A special question, on which this research can offer only very limited evidence, is whether any particular personal attributes and educational preparation are needed for those selected to play integrating roles. While further research on this subject is under way, our evidence does suggest that integrating departments could best be manned by people with prior experience in the various basic departments whose work is to be linked. Most organizations have some long-established integrating departments, even though they are not often so acknowledged. We have in mind such groups as production control, who usually work as integrators between production and sales. Some budgetary planning and cost-control people also perform this function. These roles likewise turn up in more specialized organizations—for example, some account executives in advertising agencies. Research to discover the common denominators of those individuals who perform effectively in these integrating jobs could have some very practical applications.

In summary, we should emphasize that the possibility of more systematically planning and implementing conflict resolution procedures hinges on establishing a baseline in terms of the required and actual patterns of differentiation in an organization. This baseline establishes the frame of reference for tackling the conflict resolution problem in an orderly fashion. This clarifies what the conflict is all about, what creates the differences of judgment, and what knowledge is relevant to its resolution. From this vantage point we can see why conflict must be accepted as a continuing result of living in a complex civilization. Resolution is not then put up as some final Utopian answer, but simply as a sensible solution to today's issue—with awareness that basic and legitimate differences will generate new conflicts to be resolved tomorrow. From this baseline managers can move more directly toward designing procedures and devices that are adequate for processing the flow of conflicted issues that will surely arise. We have reason to hope that the use of these guidelines can lead to a higher success rate in our experimentation with new organizational and social inventions for conflict resolution. Finally, the reader may need a reminder that an effective conflict resolution system is really the same thing as an effective decision-making system—rather central to organization performance.

Implications for the Selection of Discrete Management Practices

In referring to discrete management practices, we have in mind such things as payment systems, control systems, manpower selection, placement and promotion systems. This study's most general implication for the choice of such management practices is that it be made with thoughtful attention to the task and environment of the organizational unit affected. This point seems obvious, but those who design and select procedural systems are often at headquarters, where the convenience of using uniform practices throughout the firm often seems more important than improving the fit between the procedure and the particular job to be done. As it becomes possible systematically to tailor management practices to different tasks, a new form of consistency can be achieved. Each of the discrete practices within each major unit can be consistent with the other practices in that unit, so that all will reinforce the desired task performance. If this type of harmony can be achieved, the sacrifice of consistency in the use of any single practice across all units will be a very small price to pay. It is neither appropriate nor possible to spell out all these possibilities in detail here. We will start a short distance down this road by considering the prospects of using the basic concepts of this study in the design of control systems, evaluation and reward systems, position descriptions, and selection and placement systems.

By the term control system we mean primarily the conventional practice of accumulating data on actual events and comparing them with planned objectives. Such data can obviously be aggregated to cover different time intervals and different organizational units. Some of the basic variables we have been using could help to guide these choices in the design of control systems. Is there a sensible relationship between the time intervals used for data reporting and the time span of feedback from the environment? We have seen that, as we look at progressively higher echelons in the hierarchy, this time span tends to lengthen. Is this reflected in the control reports? We also saw that in most organizations production has a shorter time span than research. Is this reflected in the design of the control system? The degree of uncertainty of information could also be considered in control system design. Are the time interval and the detail of reporting adjusted for variations in certainty? The computer's great and growing capability for processing information makes such a flexibly designed control system an eminently practical choice.

An organization's evaluation and reward systems are usually closely tied to the control system. Some of the questions just raised concerning time span would be equally relevant to selecting the time interval to be used in formal systems for reviewing individual and group performance. Another factor of consideration in designing such systems is the degree to which each man's job requires him to play an independent, highly differentiated role or, at the other extreme, a highly interdependent integrator's role. Most jobs have some mix of these duties, but the proportions can vary considerably. Such variability, when analyzed, can help the planner to decide whether rewards are best tied to specific limited task results or to the general performance of a larger segment of the organization. It can also influence whether the linkage to results is direct or indirect. Reward systems, as we have seen, can help to induce either the "you do your job, and I'll do mine" attitude or the "let's pull together" attitude. Each can be suitable under certain conditions. Reward systems should be designed with these conditions more clearly in mind.

How much the behavior of managers is influenced by their formal position descriptions and the "standard operating procedures" handbook has always been a matter for debate. The fact of the debate is, in itself, probably a good indication that the seriousness with which these documents are taken varies from organization to organization, and even from unit to unit within any one organization. This well-known variation may occur because those who draft these ground rules do not take sufficient account of the differences in task and environment faced by each major unit or by different total organizations. We have seen in this study that it is useful to vary the degree of structure in units as the certainty of their task varies. One indication of structure is the specificity of position descriptions. Another is the degree of reliance on detailed formal rules. These findings can be directly applied as guidelines in the design of job descriptions and procedural manuals. The more certain and predictable the task, the more appropriate it is to be specific and detailed in job descriptions and rules. The reverse, of course, is also true.

We have already made a case for research beyond the limited findings of this study into the relation between individuals' predispositions and their readiness to adopt required mental orientations and live comfortably within requisite structures. Until we have hard findings from such research, we cannot say much of practical import on the subject of personnel selection, except to emphasize that a clearer picture of the desired attributes for each part of the organization provides the personnel decision maker with very useful mental guidelines for reviewing candidates. In business generally the personal interview is still by all odds the single most important selection tool. This means that the selection criteria are buried in the intuition of the interviewer. His intuition can profitably be guided by more systematic knowledge of the requisites for different organizational segments, even though we cannot now completely translate them into precise personality traits nor accurately measure these traits in candidates.

The treatment we have given the choice of management practices should properly be seen as only suggestive of the potential implications of contingency organization theory in this area. Much of the possible application work must clearly await further applied research, but the prospects for such research have not been so favorable for many years. For the past decade or so systematic inquiry into discrete management practices has been rather out of style, not only because such research is often considered "lowbrow applied stuff," but more importantly, because the few studies undertaken have proved to be of very limited value. Researchers have realized that simple surveys of existing practice add little to useful knowledge, yet more sophisticated research designs all too frequently failed to come up with meaningful findings. The trouble has been that most attempts to relate discrete management practices to measures of performance have only served to prove that the search for universally effective practices is futile. In such circumstances backing off from this research is, in the short run, the only intelligent response. Now, however, the prospects for renewing such efforts are much brighter, provided investigators control for relevant contingent variables. Such research seems to be picking up momentum, and the manager can reasonably expect some specific help from the findings that emerge.

Implications for Designing Multi-Industry and Multinational Companies

In this study we have not directly addressed the particular organizational issues raised by the added complexity of operating either in a number of different industries or in a number of different national cultures. In the past few years, however, these issues have become of vital concern to an increasing number of organizations. Therefore, even though these areas need much specific research attention, it seems useful to explain how the general concepts developed out of this research might relate to them.

The methods that we have found useful in categorizing the environmental differences among the plastics, food, and container industries could help a multi-industry company to get a sharper fix on important attributes of its various industry environments. Are all the environments heterogeneous and dynamic, like that of plastics, or are they also involved with more stable and homogeneous settings, like the containers environment? Mapping out industry differences in these terms can produce some guidelines for helping to answer such essential design questions as: Is it possible for a single basic research unit at the corporate level to be so structured and oriented that it can adequately serve all the various industrial divisions, or must we establish division-level units that can be differentially tailored to the scientific environment of each industry? Of the many technical support groups, such as legal and public relations, which can reasonably be expected to serve all divisions from a corporate-level office, and which, because of industry differences, must be established at the division level? If the number of divisions indicates the need for a level of group executives, can the systematic analysis of industry differences provide a sound basis for clustering divisions into groups? The answers to these questions will, of course, be influenced by considerations other than those of this study, but our findings suggest that the environmental attributes we have identified warrant serious attention.

The issue of integration between industry or product divisions can also be reviewed in the light of the findings of this study. One of the most troublesome problems in many multidivisional companies is that of realizing to any reasonable extent the potential payout of a free flow of ideas, methods, and resources among divisions. Each division tends to "reinvent the wheel," or to move in a direction that may not be in the interests of its sister divisions. The clarity with which we can now see how differences in structure and orientation affect the quality of integration among units helps to illuminate this issue. An analysis of the relationship among product divisions similar to the one this study makes for functional units could suggest the more promising places to establish interdivisional liaison. And our findings about the partial determinants of effective conflict resolution can suggest practices and procedures that will help to close the gaps among divisions.

The multinational company has attempted not only to span the necessary differences among functional and product specialists, but also to bridge unavoidable differences among cultures. Every society tends to inculcate its young with certain basic ideas about the "right" way to bargain, to solve problems, to reward and punish; the "right" way to show anger, trust, suspicion, respect, pleasure, and disappointment; the "right" way to keep an appointment, to honor a verbal agreement, to distinguish honest from dishonest business practices. All these and many other basic norms of everyday behavior are brought into the organization by every person in it. When these standards are different, as they must inevitably be in multinational companies, the consequences are bound to be considerable. Perhaps the concepts of differentiation and integration can help managers to think more clearly about this difficult issue. Questions that might lead to its crux are: What are the necessary interdependencies among units employing members with different cultural origins? How great is the cultural differentiation? Does this suggest the best spot to position those rare people with highly developed bicultural skills who can act as integrators?

Implications for the Process of Organizational Growth and Change

Another major question that was not directly addressed in this study is the issue of organizational growth and change. Healthy organizations are always under pressure to grow. This pressure may come from outside the system, in the form of requests to undertake new or expanded tasks, or it may come from inside, from members who seek a wider scope for their own activities and who see new opportunities to be developed. These pressures create the need to make strategic choices about the direction and rate of growth. Many factors enter into these choices.[8] The concepts of differentiation and integration can be one tool to aid in making these decisions rationally. They can help executives to understand the nature of their existing organization in terms of its current pattern of differentiation. This can be mentally tested for accuracy of fit with the proposed growth, with its implications for creating newly differentiated parts. Does the proposal set up a need for a new technology? If so, will it require structures and orientations that are similar to or different from existing patterns? For example, Woodward's study suggests that managers should be particularly alert to the organizational implications of moving from a base in one technology (*e.g.*, process) into an additional one (*e.g.*, mass production). Will the domain of the organization, as defined by its market, be changed in important ways? Managers are accustomed to watching for important changes in the physical size of their market domain, but they are not so often alert to the organizational implications of market changes. Will the new domain involve any major shift in the degree of certainty? If so, it will probably also require an organizational change that must be faced up to in advance.

Many growing companies become involved to some extent in mergers and acquisitions. The assimilation of new acquisitions is at present very uneven, marked by notable failures as well as successes. While there is little hard evidence on this matter, it would appear from anecdotal accounts that many failures result from too little attention to the organizational issues connected with acquisitions. The concepts and measurement tools of this study can contribute to the better understanding of these issues. They would be useful in determining the environmental requisites for the new acquisition. Similarly, they could provide an understanding of the actual differences in structure and orientation between the acquired company and the parent, and whether these differences should be maintained or diminished. Finally, these concepts and measurements could shed light on appropriate methods of conflict resolution. The organizational information thus provided could save a parent company from inadvertently drifting into "converting" a new acquisition into its own image, rather than allowing it to maintain its own necessary organizational practices and orientations. Too often the conversion process results in dissipating the most important asset acquired—a strong, viable organization.

Every organization, whether growing or not, is periodically faced with the necessity of bringing about some fundamental changes in the behavior of its members if it is to stay effectively related to its changing environment. Various research studies on effecting desired behavior changes in organizations have emphasized the importance of using both structural modification and education.[9] The educational approach gives people a chance to become familiar with the proposed change, to comprehend the reasons behind it, possibly to contribute to its design, and to test out behaving in new and different ways. The structural approach sets up mechanisms that serve to reward the desired behavior and punish conduct that is no longer approved. Both approaches are based on well-established learning theories, and each can serve to strengthen the other. The findings of the present research, however, can provide some guides to the sequence and emphasis that might be given these two approaches.

If management has concluded, on the basis of the type of analysis we have been suggesting, that a given organizational unit needs to be more highly structured, it would be directionally consistent to initiate such a change with a structural and procedural change carried out by the appropriate authorities. This first step could then be followed up by an educational effort to make the change more understandable and livable for the people involved. This sequence would emphasize the objective of achieving more formal structure. If the desired change is in the opposite direction, toward less formality, the initial use of an educational program would be more appropriate, followed by the official shifts in structure and procedure. The sequential strategy in this case would contribute to less reliance on formal structure and again would clearly foster a consistency between the direction of the desired change and the sequence of methods.

On a more detailed level the same kind of thinking could help guide the choice between alternative forms of educational programs. For example, let us assume that a company wishes to improve its managers' ability to resolve conflict by helping them to develop their interpersonal competence. If these managers are expected to work day in and day out in a unit with relatively low formalization of structure and procedure, it would seem appropriate to employ sensitivity-training procedures, in which the entire learning process hinges on exposing the individual to a very unstructured and ambiguous situation. It also follows that in the opposite organizational circumstances, the use of more structured educational methods would be indicated. These hypothetical situations are submitted only as examples of how the direction of desired change can itself help to guide the selection of the methods employed to implement it. The findings of this and the similar studies described in the last chapter can be used to determine the desired direction of change.

A clearer understanding of the direction of desired change can also help management to choose among alternative technologies. Throughout the discussion in this chapter we have been implicitly assuming that the technology an organization employs to perform any chosen task is automatically determined by the methods available and by hard economic facts. This is not aways true. Companies can and do, by a variety of means, convert these technical constraints into choices. A given technical capability can be used in quite different ways. For example, some supermarket chains use computers to help pull decisions on manning, commodity ordering, merchandising, and so on from the stores into centralized headquarters. Other chains have used the same computer hardware as an extension of the store manager, who is regularly provided with data about his store's operations so that he can make many key business decisions himself.

Similarly, there are often two or more technically feasible ways to perform a given task. Assembly work can often be done either serially, with the product passing from one work station to the next, where the parts are stored, or concentrically, with the parts moving into a single work station, where the product remains stationary as it is gradually built up. These alternative technologies would require different structures and procedures. As a final example, organizations have many choices available about whether to make or buy components. If a company decides to emphasize a capability for performing relatively uncertain tasks, it can often choose to subcontract for a supply of components that can best be produced in a highly routine fashion. Within limits, then, even the choice of "sub-technologies" can be influenced by an overriding strategic choice about the desired direction of change —toward an organizational capability for coping with greater or less environmental certainty.

The Viable Organization of the Future

We have considered some of the practical and shorter-term implications of this study for those with leadership responsibilities in complex organizations. However, as we indicated at the outset, one of our purposes in undertaking this study was to learn what characteristics would be required of the viable organization of the future, not just in the present. In exploring this topic we should point out that while our data and our application examples have been drawn from business settings, the implications, in a more tentative way, can be extrapolated for consideration in other complex institutions of a modern society.

The broadest implication of this and other current research

is that the behavioral sciences can now be more extensively and systematically brought into the designing of complex organizations. If this comes to pass, what will be its impact on organizations? What will a more scientifically designed organization be like? As we look at the characteristics of the viable organization, suggested by this study, we will operate on the assumption that tomorrow's organizations will be more scientifically designed for the achievement of their manifold purposes. Since American business firms have a history of relatively quickly applying any new knowledge relevant to their operations, this assumption must be taken seriously. Before proceeding into this speculation about the future, however, we need to review and expand two additional assumptions that are made earlier about environmental trends.

Environmental Trends—Faster Change and Greater Heterogeneity

The vast majority of serious commentators on the subject are agreed that the rate of scientific advance is increasing. While a few voices question this, there is no doubt about the growth rate of the numbers of people doing scientific work and about the proliferation of areas in which man is exploring the unknown. The phrase "pushing back the frontiers of knowledge" has become stale by repetition, but it is still accurate. Man does invade the unknown. Figure IX–8 helps to capture this idea as it occurs.

As we suggested in Chapter I, in considering the organization of the future we will be assuming an accelerating advance of scientific knowledge. We will also be assuming that the forms of human work will be increasingly diverse. For example, the only organizations that Udy could find to study in his nonindustrial societies were engaged in a few basic pursuits—fishing, hunting, land tillage, construction, etc.[10] Today we consider people "gainfully employed" in a vast ar-

Figure IX–8

DEVELOPMENT OF KNOWLEDGE

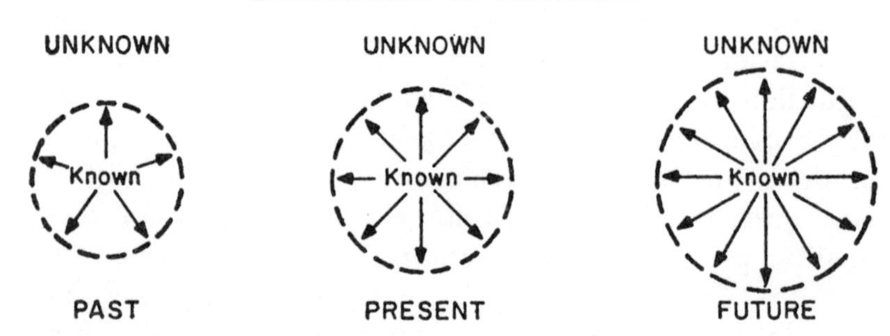

ray of different activities. Man is constantly dreaming up new activities that become generally defined as "work" as soon as others are willing to reward him for performing them. One of the fundamental reasons for the increasing diversity of work is, of course, that man has invented machines that can produce most of his survival commodities, and has thereby freed himself to invent new forms of "work." This becomes clearer if we think of the sources of all productivity as broken into four sectors: (1) all-machine work; (2) man-machine work (in which man uses himself as a guide and feeder to a machine); (3) man-tool work (in which man guides and powers a simple tool); and (4) all-human work. Thinking in these terms, we can ask how important each sector has been as a source of gross productivity as man moved from a nonindustrial into an industrial society. We know of no way to measure these matters precisely, but Figure IX–9 is a schematic attempt to represent the probable gross trends.

In this diagram we have also indicated by dotted lines the process by which new conceptualizations of the unknown are carried through the multiple stages of uncertainty-reduction to the point where machines can be completely programmed to do the work. By the spacing and slope of these lines we have indicated that today this process is happening more fre-

Figure IX–9

SOURCES OF PRODUCTIVITY

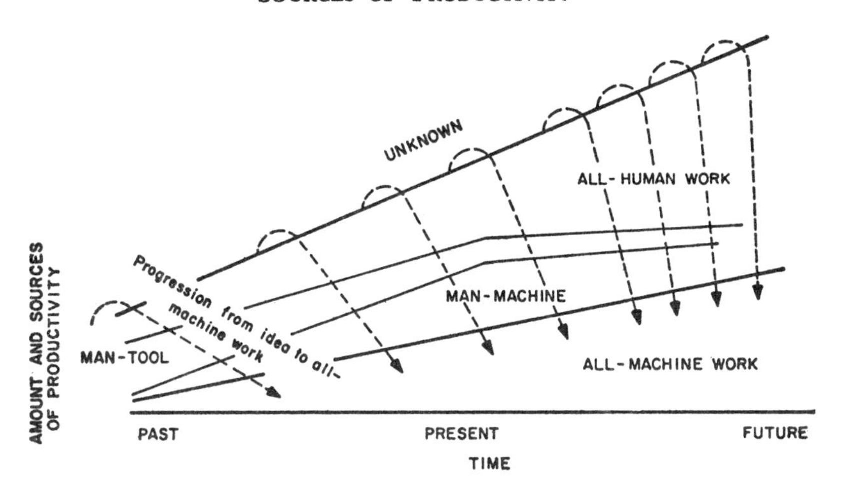

quently and with a shorter time cycle, and that this trend will continue into the future.

A chart of this kind does not, of course, prove anything. It can, however, help us to question some frequently made assumptions about the impact of the modern industrialization process. With all the current discussion of the growth of automation (the all-machine sector), it reminds us that the "all-human sector" is also growing, not only absolutely but relative to the man-tool and the man-machine sectors. It reminds us also of the great and increasing diversity of "all-human" work. One major category of all-human work is the transformation of ideas into all-machine work. This corresponds roughly to the work of R&D departments in business organizations. Another is the management of all-machine work. The growth of work of this administrative type is seldom noted by the public. A third rapid growth category, in the "all-human" sector, involves the helping of other humans. This includes, among many other things, the work of identifying, developing, and gratifying newly emerging human needs—the work of marketing and sales departments. Above all, the chart can help us to see the validity of the assumption that the discrete tasks in which man will be engaged in the future will increase not only in number but in heterogeneity. As the comparison of the three industries in this study suggested, the industrial environment of the future will be both less certain and more diverse. What will be the character of the viable organization in such an environment?

Implications for the Form and Size of Viable Organizations

If our projection of environmental trends proves accurate, the viable organization of the future will need to establish and integrate the work of organization units that can cope with even more varied sub-environments. The differentiation of these units will be more extreme. Concurrently, the problems of integration will be more complex. Great ingenuity will be needed to evolve new kinds of integrative methods. The viable organizations will be the ones that master the science and art of organization design to achieve both high differentiation and high integration. The high-performing plastics organizations offer at least one model of the direction in which such organizations may evolve. To conceive new organizational forms and to develop the managerial behavior needed in these viable organizations, managements will rely increasingly on formally designated organizational development departments. These departments, staffed by trained specialists in the behavioral and administrative sciences, will be involved in planning new organizational forms for the effective utilization of human resources and in training managers to operate effectively in these settings. As John W. Gardner suggests:

> What may be most in need of innovation is the corporation itself. Perhaps what every corporation (and every other organization) needs is a department of continuous renewal that

> could view the whole organization as a system in need of continuing innovation.[11]

As the demands for both differentiation and integration become more acute, top management will also find it necessary to devote more and more of its explicit attention to the achievement of these organizational states.

The organizations that achieve these states of differentiation and integration will grow rapidly in size. They will do this simply because they will be able to perform their various purposes more effectively than separate smaller organizations. The factors we believe will contribute to the growth of these organizations can usefully be analyzed with the economic concepts of "vertical and horizontal integration."

Vertical integration involves acquiring organizational control and property rights over the sources of semifinished and raw materials, or over distribution channels to the ultimate consumer. It is concerned with a single flow of materials and services from raw states to consumable form within one corporate entity. Such consolidations have often been initiated for defensive reasons—to reduce uncertainties about access to raw material sources and to markets. For whatever reason they occur, they have had the indirect effect of removing certain transactions of goods and services from the open marketplace and placing them within the purview of a single organization. The marketplace is, of course, one type of integrating device, and consolidations that convert marketplace transactions into intraorganizational transactions will not be viable over time unless the intraorganizational integrating devices prove more effective than the marketplace. With more sophistication in the design and use of integrating devices, more such vertical consolidations will be warranted. The limits on what can be done efficiently within a single organization will be pushed back. Thus the viable organization of the future will be a growing organization.

A somewhat similar trend will mark growth by horizontal

integration. This is the process by which companies diversify into new product lines, becoming, in the current phrase, "conglomerate." Again, these expansions are often initiated for defensive reasons—spreading of risks, shifting from a declining to an expanding product line, etc. Such an expansion, however, is likely not to be viable over time unless the organization creates among these diverse product lines a flow of ideas and methods superior to their dissemination among separate organizations. Horizontal integration will not be warranted unless a synergistic effect is achieved. Again it comes back to the organization's ability to create superior integrating methods—in this instance, for the exchange of help among product divisions. As the capacity for solving these organizational issues increases, horizontal expansion will be fostered. Again the limits of what can be done more efficiently within a single organization will be pushed back, and organizations with this capability will expand.

The lines of growth examined here are already historical realities. Increased organizational capability will simply carry these existing trends much further. The traditional lines dividing industries will be further eroded. The original national identity of corporations will be further obscured. Competition will more and more be directly among *all* of these multipurpose corporations, regardless of their industry or national origins.

Characteristics of Viable Organizations

The viable organizations we are projecting for the future will have reasonably mastered the ability to organize work that ranges from basic scientific endeavors to highly standardized and routine production. They will have learned to operate all over the globe and to link these operations together. They will be able to move repeatedly into new product lines with a sure-handed grasp of markets and technologies. They will, in this way, be able to organize effort toward the achievement of human purposes that are not now even conceived. Such "multi-organizations" would be able to undertake and effectively perform assignments in what is currently defined as the public, as well as the private, sector. They could engage in tasks that are now impossible to perform.

The process of creating such potent organizations will generate a good deal of experimentation and ferment. We will probably see a period in which our rapid technological change is matched by rapid organizational change. After these organizations have taken shape as highly differentiated and highly integrated entities, however, the pace of organizational change can be expected to slacken, even while these same organizations speed up the rate of technical change.

For the individual, life in these multi-organizations will not necessarily resemble the stereotype of the grey-flanneled organization man in the crystal palace. Instead the great diversity of required roles can give meaningful scope to the potentialities and career aspirations of a fairly wide variety of people. Procedures should be more effective in helping people find the succession of assignments that meet their developing needs and personal abilities. The organization will serve as a mediator or buffer between the individual and the full raw impact of technological change by providing continuing educational opportunities and various career choices. The multi-organization will be highly rationalized, but not in the old-fashioned sense of everything being programmed into rules. The reliance on detailed programming will vary greatly among the different parts of the organization. The differentiation of parts of the organization will guarantee the continuing existence of conflict, but the aura of rationalization of the new type can make these conflicts less personalized. Since more diverse opportunities will be available, there will be less justification for strictly person-to-person fighting for career survival, with its brutality and human waste.

We may find a division of labor based more on cultural differences emerging in these organizations. The different countries have traditionally divided labor partially on the basis of control of raw materials and, more recently, on technical and organizational capabilities. These sources of an international division of labor may be supplemented in the future by one based on differences in values. Basic cultural values are among the slowest-changing aspects of human life. These values prepare people to play some occupational roles better than others. Perhaps our multi-organizations will be able to build on these differences, to design a division of labor around them, and to reduce the present strong trend toward a universal culture of industrialized man.

We have been presenting the ways in which the emergence of multi-organizations could make it possible to achieve results that most people would consider desirable. Clearly, these organizations also have potentials for evil. In many current works of literature and art the organization is cast as the villain that oppresses the free spirit of the individual. There are undoubtedly certain real dangers that the development of multi-organizations could result in a greater concentration of power in a few over the working lives of the many. One important guarantee against such misuse of the multi-organization seems to lie in the study and understanding of their functioning. Such understanding can contribute to an informed public opinion, which is the foundation of any effective countervailing power, whether it be channeled through labor unions, government, or some new institutional arrangement.

Perhaps another important countervailing power resides within the organization—its leadership. Too often those who criticize existing organizations and are concerned about their constraining effects on individuals seem to suggest that organizations have a will and purpose of their own. Even the best-designed organization, now and in the future, is but a tool, albeit a complex one, for the achievement of human purposes. Organizations are and always will be run by people. They are infused with purpose and meaning only through the imagination and will of people—by acts of leadership. As Chandler emphasized, organization design and structure must follow from and be subservient to strategy—human purposes formulated into organizational goals.[12] It is by these revitalizing acts of leadership that organizations remain useful tools, not stultifying masters. Leadership, then, is a crucial issue, because it provides one safeguard against the risks of multi-organization. Since this book has been primarily addressed to the managers who are or will be leaders of these organizations, we shall conclude by examining the implications of this study for leadership behavior in the effective organization of the future.

Leadership and Multi-Organizations

The viable multi-organization, like the effective organizations of today, will require strong leadership to provide direction to the enterprise. Unlike many of today's effective organizations (*e.g.*, the high-performing container organization), however, the effective multi-organization will require not just a few such leaders, but many of them. All the evidence of this and other research points to the need for multiple leadership in these complex organizations. As they cope with heterogeneous and dynamic environments, the issues and knowledge involved become too complicated for only a few leaders to understand. Simpler organizations dealing with more homogeneous and stable environments can be led by a few great men, and this gives these leaders great prominence. They fit the picture of the romantic hero. The fact that the multi-organization needs many leaders may reduce the unique prominence of any one of them, but this must not obscure the importance of each. All these leaders will have to face the tensions of making commitments of great consequence. These will be the modern heroes, even if they are not conspicuous to the general public.

Perhaps even more importantly, several different leadership roles will be required in the multi-organization. We saw this in the high-performing plastics firm. Those parts of the organization that are centers of innovation and entrepreneurship must be led by managers with the capacity to innovate and take risks. These leaders will have to be skilled in dealing with the unpredictable and ambiguous scientific and market sectors of the environment. They will require the capacity to deal effectively with creative scientific, technical, and marketing specialists. Those parts of the organization performing more routine tasks will require different kinds of leaders who will fit more closely our traditional leadership models. These men will be needed to lead the specialists performing the more programmed tasks that are more likely to appear in production and sales areas.

The need for these two types of leadership for the multi-organization is not particularly novel. Managers have for some time been aware of the requirements for both of these roles. The leadership role that does seem to be unique to the multi-organization is that of the integrator, who must actively link together its many parts. Too often in the past the term "coordinator" has conjured up images of a passive, responsive individual who transmitted information back and forth between more active managers. The findings of this study clearly suggest that this stereotype does not apply to integrators in effective multi-organizations. These people will need to have high influence in decision making, based on their unique perspective. They will need to be leaders who are clearly identified with and committed to the success of their product groups. They will have to be leaders who can take the initiative in setting goals and who have the interpersonal skills to achieve the resolution of difficult conflicts.

Finally, the multi-organization at its top level will require leadership that can formulate a general framework of purpose to guide the efforts of the parts. This will require the highest order of integrative and creative capacity. Perhaps one of the most important functions of these top managers will be the designing of new forms of complex organizations to better achieve the multiple purposes of our evolving civilization. In essence, their problem will be to create and use multiple centers and multiple styles of leadership. In the organization of the future, acts of individual leadership will not be out of date. To the contrary, these acts, given the increased environmental ambiguity and the increased freedom to choose organizational ends and means, will be more crucial than ever in determining organizational health and, beyond that, the well-being of our human society.

Methodological Appendix

The material which follows is intended to give the interested reader a description of the questions asked in questionnaires and in interviews. The more complicated analytic procedures are also described. This material is organized according to the major variables we have attempted to measure. The reader who is interested in even further details on our research and analytic methods is referred to three sources:

Paul R. Lawrence, and Jay W. Lorsch, "Differentiation and Integration in Complex Organizations," *Administrative Science Quarterly,* June 1967.

Jay W. Lorsch, *Product Innovation and Organization* (New York: The Macmillan Company, 1965).

James S. Garrison, "Organizational Patterns and Industrial Environments," unpublished doctoral dissertation, Harvard University Graduate School of Business Administration, 1966.

Measures of Environmental Demands

Environmental Interview Questions

The questions below provided the structure for extensive interviews with the top executives (chief executive and the head of each functional department) in each organization.

1. As a place to start, on what basis does a customer evaluate and choose between competing suppliers in this industry? (Price, quality, delivery, service, etc.)
2. Would you list for me the major kinds of problems an organization encounters competing in this industry?
3. Would you rank these problem areas in terms of: (a) their criticalness to the success of the organization; (b) the difficulty of achieving effective resolution?
4. How do you go about solving these most critical problems?

5. To what degree are you comfortable in dealing with these problems, given the nature of the information available to you? For example, how would you characterize your conviction (high, so-so, low) in specifying the:

   (a) appropriate research or design activity?
   (b) appropriate processing procedures?
   (c) appropriate product specifications and sales and service activity?
   (d) appropriate activity to counter competitors' abilities?

6. Assuming for the moment that you could make a limited number of requests for any additional information you desire, what information would you seek?
7. We have talked a little about the basis on which products of different companies in this industry tend to compete. I would like your impression of how important each functional area is in determining the final product characteristics?
8. To what extent (considerable, some, very little) must most product characteristics be worked out interdepartmentally, in contrast to one functional unit being able to determine the characteristics required and achievable and then passing them on to the other unit? What interdepartmental relationship would you label most critical in this industry?
9. Have there been any significant changes in the market or technical conditions in your industry in the past 10 years? Which have been the most important to your organization?
10. To what extent have there been major modifications in the following activities in your division over the past 10 years:

    (a) in your product line?
    (b) in your marketing techniques?
    (c) in your manufacturing facilities?
    (d) in the amount and direction of your research effort?
    (e) in the background, training, and technical skills of your employees—sales, manufacturing, research, management?

Environmental Questionnaire

These questions were asked of top executives in each organization. Terminology was adapted slightly to particular industry and company circumstances.

Questions pertaining to environmental certainty:

> Due to rapid change in an industry, or the state of development in the technology used by the industry, or vast differences in customer requirements, etc., company executives often have varying degrees of certainty concerning what their departmental job requirements

are and the kinds of activities their departments *must* engage in to achieve these requirements. The following series of questions is an effort to obtain data concerning this aspect of your industry. Please answer each question for each functional area.

(a) Please circle the point on the scale provided which most nearly describes the degree to which present job requirements in each functional department are clearly stated or known in your company for the:

Research Department

Job requirements are very clear in most instances	1 2 3 4 5 6 7	Job requirements are not at all clear in most instances

Manufacturing Department

Job requirements are not at all clear in most instances	1 2 3 4 5 6 7	Job requirements are very clear in most instances

Marketing Department

Job requirements are very clear in most instances	1 2 3 4 5 6 7	Job requirements are not at all clear in most instances

(b) Please circle the point on the scale provided which most nearly describes the degree of difficulty each functional department has in accomplishing its assigned job, given the limitation of the technical and economic resources which are available to it:

Degree of difficulty in:

Developing a product which can be manufactured and sold profitably

1 2 3 4 5 6 7
Little difficulty — Extremely difficult

Manufacturing economically a product which can be designed and sold

1 2 3 4 5 6 7
Extremely difficult — Little difficulty

Selling a product which can be developed and manufactured economically

1 2 3 4 5 6 7
Little difficulty — Extremely difficult

(c) Please check the alternative which most nearly describes the typical length of time involved before feedback is available to each functional area concerning the success of its job performance. For example: the sales department manager may be able to determine at the end of each day how successful the selling effort was by examining the total sales reported by his salesmen for that day. In contrast, the production manager may not know whether production meets required specifications until the results of several performance tests are available, often a period of several days from the time his department completes its processing.

Research Department

______________________ (1) one day
______________________ (2) one week
______________________ (3) one month
______________________ (4) six months
______________________ (5) one year
______________________ (6) three years or more

Manufacturing Department

______________________ (1) one day
______________________ (2) one week
______________________ (3) one month
______________________ (4) six months
______________________ (5) one year or more
______________________ (6) three years or more

Marketing Department

______________________ (1) one day
______________________ (2) one week
______________________ (3) one month
______________________ (4) six months
______________________ (5) one year
______________________ (6) three years or more

Questions pertaining to required integration:

We are interested in the degree of coordination required between the various functional departments in a typical company in your industry. The following series of questions is concerned with this aspect of your industry.

(a) We would like you to check the statement which most nearly describes the extent to which each functional area is able to define its job objectives and to make changes in its activities on its own.

Research:

To an extreme extent	To a very great extent	To a considerable extent	To some extent	To a small extent	To very little extent	Not at all
______	______	______	______	______	______	______

Manufacturing:

Not at all	To very little extent	To a small extent	To some extent	To a considerable extent	To a very great extent	To an extreme extent
______	______	______	______	______	______	______

Marketing:

To an extreme extent	To a very great extent	To a considerable extent	To some extent	To a small extent	To very little extent	Not at all
______	______	______	______	______	______	______

(b) We would like you to check the statement which most nearly describes the extent to which the *Research Department* is influenced by how:

Manufacturing performs its task

To an extreme extent	To a very great extent	To a considerable extent	To some extent	To a small extent	To very little extent	Not at all
______	______	______	______	______	______	______

Marketing performs its task

To an extreme extent	To a very great extent	To a considerable extent	To some extent	To a small extent	To very little extent	Not at all
______	______	______	______	______	______	______

(c) We would like you to check the point which most nearly describes the extent to which the *Manufacturing Department* is influenced by how:

Research performs its task

To an extreme extent	To a very great extent	To a considerable extent	To some extent	To a small extent	To very little extent	at Not all
______	______	______	______	______	______	______

Marketing performs its task

To an extreme extent	To a very great extent	To a considerable extent	To some extent	To a small extent	To very little extent	Not at all
______	______	______	______	______	______	______

(d) We would like you to check the point which most nearly describes the extent to which the *Marketing Department* is influenced by how:

Research performs its task

To an extreme extent	To a very great extent	To a considerable extent	To some extent	To a small extent	To very little extent	Not at all
______	______	______	______	______	______	______

Manufacturing performs its tasks

Not at all	To very little extent	To a small extent	To some extent	To a considerable extent	To a very great extent	To an extreme extent
______	______	______	______	______	______	______

Question pertaining to the relative importance of parts of the environment:

Below is a list of the major functional specializations involved in your industry, developed by examining the organization charts of a number of firms in the industry. While an adequate performance by each of these departments is certainly necessary for company survival, a high level of competence in one or two of these departments may be more critical to your success than others in this industry. We would like you to rank the departments listed below in terms of the importance of each in contributing to your company's ability to compete successfully in this industry.

Place a "1" beside the department you feel to be most critical
Place a "2" beside the department you feel to be next most critical

______________ Research
______________ Sales
______________ Manufacturing

Computation of Required Differentiation and Required Departmental Attributes

Required departmental attributes (Chapters II and IV) were determined based on predictions of the orientations and structure necessary to deal effectively with the parts of the three environments, and this was compared with actual attributes. For example, in the plastics industry the following analysis was made to determine whether the units met the demands of their part of the environment:

> In regard to structure, we trichotomized the structure scores for all units. Fundamental research units were expected to fall in the lowest third (less structured), sales in the middle third, and production in the highest third. The interpersonal orientation scores were dichotomized: production and fundamental research were expected to fall in the task half of the range, sales in the social half. As a test of whether the time and goal orientations of a unit were consistent with the demands of its part of the environment, units in the lowest third of the range in orientation toward required time and goal dimensions were defined as not effecting a satisfactory fit.
>
> The number of deviations from these environmental demands for organization in each performance category was then calculated. In this analysis it was necessary to exclude the applied research units because of their variable functions in the plastics organizations.
>
> A similar analysis was made for the container and food industries.

Below is an example (from the container environment) of the calculations used to determine the required differentiation scores (Chapter IV).

The raw scores for each part of the environment were as given in Table A–1.

TABLE A–1

	Environmental Sector		
	Science	*Market*	*Techno-economic*
Uncertainty score	7.4	5.8	7.8
Time span of definitive feedback	3.4	2.6	2.1
Relative importance of environmental sectors	2.4	1.8	1.8

The next step was to compute the differences between each sub-unit score for each dimension. The differences along the time span and certainty dimensions, which were used to determine required differentiation in time orientation and in structure and interpersonal orientation respectively, were computed by straight subtraction. These differences are shown in Table A–2.

The required differences in goal orientations were calculated from the relative importance of environmental sector scores. The primary orientation of each unit's membership was expected to be directed toward the concerns of their own part of the environment. The existence

TABLE A–2

Sub-unit pairs	*Differential* *Certainty*	*Time Span*
Research—Marketing/Sales	1.6	0.8
Research—Production	0.4	1.3
Production—Marketing/Sales	2.0	0.5

of a dominant part of the environment was expected to affect each of the unit's secondary concerns, skewing their total orientation toward the dominant portion of the organization's environment. The net result would be that large differences in the relative importance of the three sub-environments would act to *reduce* the degree of required differentiation between the sub-unit's goal orientations. Consequently, the degree of required differentiation in goal orientation was computed as follows:

First the required goal orientations for each unit were ranked in accordance with the ranking of the relative importance of each part of the environment (Table A–3).

TABLE A–3

Environmental sector	*Unit* *Research*	*Sales*	*Production*
Scientific	2	3	3
Market	3	1	2
Techno-economic	1	2	1

Next, the differences in these rankings in all three goal orientations were calculated for each pair of sub-units; with the results shown in Table A–4.

TABLE A–4

Sub-unit pairs	*Ranking*
Research—Sales	4
Research—Production	2
Production—Sales	2

Finally, the differences in rankings for each pair of units was *reduced* by the differences in the relative importance of their part of the environment (see Table A–1 above), to arrive at a final index of the required degree of differentiation in goal orientation, as given in Table A–5.

TABLE A–5

Sub-unit pairs	*Computation*
Research–Sales	4 – 0.6 = 3.4
Research–Production	2 – 0.6 = 1.4
Production–Sales	2 – 0.0 = 2.0

Having computed these required differences for each pair of units along three dimensions, the range of these differences for all pairs in the three environments was divided into quintiles and each quintile was assigned a "required differentiation" score. These scores and the range of differences are presented in Table A–6.

TABLE A–6

	Score				
Dimension	*1*	*2*	*3*	*4*	*5*
Task	1.20–1.64	1.65–2.09	2.10–2.54	2.55–3.00	3.01–3.44
Time span	0.10–0.52	0.53–0.94	0.95–1.36	1.37–1.78	1.79–2.20
Certainty	0.20–1.34	1.35–2.49	2.50–3.64	3.65–4.78	4.79–5.92

The raw differential scores (in parentheses) for each pair of units in the container industry and the required differentiation score (as presented in Table IV–4) is shown in Table A–7.

TABLE A–7

	Required Differentiation			
Sub-unit pairs	*Task*	*Time Span*	*Certainty*	*Total*
Research–Marketing/Sales	5	2	2	9
	(3.4)	(0.8)	(1.6)	
Research–Production	1	3	1	5
	(1.4)	(1.3)	(0.4)	
Production–Marketing/Sales	2	1	2	5
	(2.0)	(0.5)	(2.0)	

Measures of Actual Departmental Attributes

Structure

To measure the structure of the departments, dimensions suggested by Hall, Woodward, Evan, and Burns and Stalker that could be operationally measured were used: the span of supervisory control, number of levels to a supervisor shared with other departments, the specificity of review of department performance, the frequency of review of department performance, the specificity of review of individual performance, and the emphasis on formal rules and procedures. See R. H. Hall, "Intraorganizational Structural Variables," *Administrative*

Science Quarterly, December 1962; Joan Woodward, *Management and Technology* (London: Her Majesty's Printing Office, 1958); William Evan, "Indices of Hierarchical Structure of Industrial Organizations," *Management Sciences,* September 1963; Tom Burns and G. M. Stalker, *The Management of Innovation* (London: Tavistock Publications, 1961). Data on these characteristics for each department were gathered from organizational documents (organization charts, procedural manuals, and the like), or when these were not available, by interviewing the department manager about organizational practices.

A four-point scale was developed for each structural characteristic (see Table A–8), and a structural score was computed for each department in all organizations by adding scores on all six characteristics.

TABLE A–8

Structural Characteristics	*1*	*2*	*3*	*4*
Average span of control	11–10 persons	9–8 persons	7–6 persons	5–3 persons
Number of levels to a shared superior	7 levels	8–9 levels	10–11 levels	12 levels
Time span of review of departmental performance [a]	Less than once each month	Monthly	Weekly	Daily
Specificity of review of departmental performance	General oral review	General written review	One or more general statistics	Detailed Statistics
Importance of formal rules	No rules	Rules on minor routine procedures	Comprehensive rules on routine procedures and/or limited rules on operations	Comprehensive rules on all routine procedures and operations
Specificity of criteria for evaluation of role occupants	No formal evaluation	Formal evaluation—no fixed criteria	Formal evaluation—less than 5 criteria	Formal evaluation—detailed criteria—more than 5

[a] Based on shortest review period.

Interpersonal Orientation

"The Least Preferred Coworker" instrument developed by Fred E. Fiedler was utilized to measure this dimension. For a description of it see *Technical Report No. 10,* Group Effectiveness Research Laboratory, Department of Psychology, University of Illinois, May 1962 or Fiedler's more recent book, *A Theory of Leadership Effectiveness* (New York: McGraw-Hill Book Company, 1967).

Time Orientation

The following question measured time orientation:

Persons working on different activities are concerned to differing degrees with current and future problems. We are interested in learning how your time is divided between activities which will have an immediate effect on company profits and those which are of a longer-range nature. Indicate below what percent of your time is devoted to working on matters which will show up in the division profit and loss statement within each of the periods indicated. Your answers should total 100%.

(a) 1 month or less ________
(b) 1 month to 1 quarter ________
(c) 1 quarter to 1 year ________
(d) 1 year to 5 years ________

Items (a) and (b) were combined to arrive at the short-term orientation, while (c) was used as medium range and (d) was used as long range.

Goal Orientation

The wording of the question used to measure goal orientation was varied slightly depending upon the dominant issue in each industry. The example below was used in the plastics industry. The part of the environment each item pertained to is indicated in parentheses, but was not included on the questionnaire. The item referring to total company profits was not utilized in our analysis.

In evaluating and considering the potentialities of a new idea, there are many considerations about which persons in different parts of the organization must be concerned. We recognize, while all of these concerns are important, that certain concerns will be most important to you. In order to learn which are most important in your personal opinion, we would like you to rank the ten criteria listed below as follows:

(a) Place a "1" by the three criteria which are of most concern to you personally.

(b) Place a "2" by the *next three criteria* which are of second most concern to you personally.

Criteria:

________ The manufacturing costs associated with products resulting from the proposed idea. (Techno-economic)
________ Competition's response to products resulting from the proposed idea. (Market)
________ The potentialities for scientific publication which might result from the proposed idea. (Science)
________ The return on investment the company might gain from the proposed idea. (Not used)
________ The technical processing problems which might result from the proposed idea. (Techno-economic)
________ The contribution which research on the proposed idea might make to scientific knowledge. (Science)
________ The capability of the sales organization to sell a product resulting from the proposed idea. (Market)
________ The technical capability of the research staff to conduct research on the proposed idea. (Science)
________ The plant facilities which would be required for a product resulting from the proposed idea. (Techno-economic)
________ The effect of products resulting from the proposed idea on the sales of existing company products. (Market)

Degree of Differentiation

To determine the relative degree of differentiation between pairs of units or among organizations it was necessary to develop a comparable differentiation score for all four attributes, i.e., structure and interpersonal, time and goal orientations. This was done as follows. First, the differences in each attribute for all ten organizations were divided into five classes. Each class was next assigned a *differentiation score* from one (least differentiated class) to five (most differentiated class). These five point units of differentiation scores for each attribute made it possible to arrive at a rough measure of the relative total differentiation between relevant units by summing the scores for all four attributes. The average differentiation scores were computed from the total differentiation scores for comparable pairs of units in each organization.

Degree of Integration

The question below was used to measure the degree of integration in each organization. The name of units was varied somewhat with each organization. The scores as originally used ranged from 1 (highest quality integration) to 7 (lowest quality). However, for ease of interpretation for the reader, these scores were reversed in our presentation. Thus 7 indicates the highest quality of integration and 1 the lowest. While a seven-point scale was used, the respondents generally used only the upper part of this scale (those items indicating higher quality integration). It was possible to check the validity of responses to this question in interviews and it was found that mean scores of worse than 2.6 (as obtained in the questionnaire or 5.4 as presented for the reader) for a pair of units seemed to indicate that there were appreciable difficulties in achieving integration.

We would like to know about relationships between different parts of the organization. This question is aimed at obtaining your evaluation of the relations between various units.

Listed below are eight descriptive statements. Each of these might be thought of as describing the general state of the relationship between various units. We would like you to select that statement which *you feel* is most descriptive of *each* of the departmental relationships shown on the grid and to enter the corresponding number in the appropriate square.

We realize you may not be directly involved in *all* of the departmental relationships indicated. However, while you may lack direct involvement, you probably have impressions about the state of the relationship between the various departments listed. We therefore would like you to fill out the complete grid.

	Sales	Manufacturing	(Title of Integrating Unit)	Applied Research
Manufacturing				
(Title of Integrating Unit)				
Applied Research				
Fundamental Research				

Relations between these two units are:

1. Sound—full unity of effort is achieved.
2. Almost full unity.
3. Somewhat better than average relations.

4. Average—sound enough to get by even though there are many problems of achieving joint effort.
5. Somewhat of a breakdown in relations.
6. Almost complete breakdown in relations.
7. Couldn't be worse—bad relations—serious problems exist which are not being solved.
8. Relations are not required.

Differentiation and Integration

A rank order comparison of differentiation and integration scores for low-performing plastics organization A, which was used in Figure II-2, is given in Table A-9. These data for the other plastics organizations are available in Paul R. Lawrence, and Jay W. Lorsch, "Differentiation and Integration in Complex Organizations."

TABLE A-9[a]

	Differentiation score	*Integration score*
Integrating Department—Applied Research	6 (1)	5.24 (1)
Integrating Department—Sales	7 (2)	5.22 (2)
Integrating Department—Production	8 (3)	4.90 (3)
Production—Applied Research	9 (4)	4.88 (4)
Integrating Department—Fundamental Research	11 (5)	4.68 (5)
Sales—Applied Research	13 (6)	4.55 (6)

[a] High integration score indicates tighter integration. Higher differentiation score indicates greater differentiation. Numbers in parentheses indicate rank order of differentiation or integration score.

To compute the average differentiation and integration scores for each organization, as we suggested above it was necessary to use comparable pairs of units in all organizations. This was a problem between industries and within the plastics industry where various units had required integration with fundamental research. In the plastics industry those relationships listed in Table II-4 were utilized. For cross-industry comparisons, as indicated in Table IV-6, sales-production, production-research, and sales-research relationships were used. The research units used in the plastics organizations were the applied ones, since these were more comparable to research units in the organizations in the other two environments.

ORGANIZATIONAL PERFORMANCE

The following questionnaire was used to solicit performance information from the chief executive of each organization. For the plastics organizations information was also gathered about what percent of current sales were accounted for by products introduced in the past five years. Similarly, for the two food organizations the number of products introduced in the five years prior to the study was also obtained. In addition clinical data were gathered in interviews with the top executives about the performance of their organization.

As the final stage of the research project in which your organization has participated, we are interested in obtaining some assessment and measurement of the performance of your company (division). We recognize that the information for which we are asking is sensitive, and therefore we want to be explicit about the manner in which it will be used. The data in the form in which we are asking you to report it will only be seen by the research group and will not be published. Instead it will be used to develop rank order comparisons between the various organizations which have participated in the study.

1. *Total Organizational Performance*

   We need to obtain your subjective assessment of the performance of your entire organization as it relates to competitors in this industry. Equating 100% to ideal performance we would like you to indicate what percent of this ideal or optimal performance you personally feel the (your) organization is achieving in this industry.

   I personally feel that the overall performance of the organization for which I am responsible should be rated as _____% in the "X" industry.

2. *Empirical Measures of Performance Over Time*

   We are also interested in obtaining a few empirical measures of the trend of your organization's performance over the past five years. In the table below we would like you to indicate the percentage *change on a year-to-year basis* of three performance indicators: sales; before tax profits; and return on investment before taxes. Considering the base year 1960 (or the year five years before the study) as 100, would you please indicate, in the spaces provided below, the level for each indicator for each year. For example, if sales in 1961 were 5% above 1960, you would put 105 in the 1961 column. If sales were 5% below the 1960 level in 1962, you would put 95 in the 1962 column, and so forth.

	1960	*1961*	*1962*	*1963*	*1964*
Sales	100				
Before Tax Profits	100				
Return on Investment Before Tax	100				

Determinants of Effective Conflict Resolution

Intermediate Position of Integrators

The methods used for measuring structure, and time, goal, and interpersonal orientations were also used for this analysis. The midpoint of the range of scores in each attribute was computed for the basic departments being integrated. The difference between the score for the integrating department and the midpoint was then computed to determine how closely the integrating department approached an intermediate position. In structure and interpersonal orientation, where there was only a single mean score for each unit, this procedure was straightforward. However, since in time and goal orientation there were three dimensions to each attribute, the procedure was somewhat more complicated. In time orientation the differences in only short- and long-term orientations were considered, since these were the dimensions where the greatest differences existed in all six organizations. The differences in both dimensions were summed to get a single score. In goal orientation only those units which were concerned with a particular part of the environment were considered. For example, in orientation toward the market, only the differences between the integrating unit and sales and research were considered, since these were the units between which the integrating unit was providing a flow of marketing information. The differences between the integrating unit and the other departments in orientation toward the market, toward the scientific, and the techno-economic parts of the environment were then summed to get a single score.

The integrating units in the various organizations were then compared to determine if they had a low or high difference from the midpoint of the range of scores in each attribute.

Departmental Influence

The following question was used to measure the relative influence of each department. Changes in departmental titles were made for each organization, and the wording about the major issues was changed for each industry.

> In general how much say or influence do you feel each of the units below has on product innovation decision. Please use the scale below. You *may* use the same score to describe more than one unit.

1. Little or no influence.
2. Some influence.
3. Quite a bit of influence.
4. A great deal of influence.
5. A very great deal of influence.

Sales	____________
Manufacturing	____________
Integrating Unit	____________
Applied Research	____________
Fundamental Research	____________

Hierarchical Influence

The following question was used to measure hierarchical influence within the respondent's own department. Again the wording about the major issues was changed for each industry. Position and departmental titles were also changed to suit each organization. In computing hierarchical influence patterns (Chapters V and VI) the levels in the various departments were matched and the mean influence was computed for each level.

In general, how much say or influence do you feel each of the following groups or individuals have on product innovation decisions. Please respond *only* for *your own department or unit* using the scale provided below. You *may* use the same score to describe more than one group or position in your department.

1. Little or no influence.
2. Some influence.
3. Quite a bit of influence.
4. A great deal of influence.
5. A very great deal of influence.

Sales (To be answered by Sales Personnel only)

Sales Manager	____________
District Managers	____________
Assistant Product Managers	____________
Salesmen	____________
Product Managers	____________
Regional Managers	____________

[*Integrating Unit*] (To be answered by [Integrating] Personnel only)

Section Heads	____________
Technical Specialists	____________
Department Managers	____________
Chemists or Engineers	____________
Group Managers	____________

Manufacturing (To be answered by Manufacturing Personnel only)

Section Managers ________
Production Manager ________
Superintendents ________
Engineers ________

Applied Research Laboratory (To be answered by Applied Research Personnel only)

Laboratory Director ________
Group Leaders ________
Senior Research Chemists/Engineers ________
Research Chemists/Engineers ________
Chemists/Engineers ________

Fundamental Research Laboratories (To be answered by Fundamental Laboratory Personnel only)

Research Chemists ________
Group Leaders ________
Assistant Laboratory Director ________
Laboratory Director ________
Associate Scientists ________

Perceived Reward

The following question measured managers' perceptions of what criteria they were evaluated on and thus what they were rewarded for. The question was asked of all respondents, but only the data for integrators was analyzed.

Persons in different organizations are evaluated in different ways. The statements below list five common bases for evaluating individual performance. You are asked to choose three of these statements which best describe the basis on which you are evaluated by your superiors, and to rank them using the following scale.

1. Describes the most important basis for evaluation.
2. Describes the next most important basis.
3. Describes the third most important basis.

_____ I am evaluated by my superiors on the basis of the performance of my subordinates.
_____ I am evaluated by my superiors on the basis of overall performance of the product group with which I am working.
_____ I am evaluated by my superiors on the basis of my own individual accomplishments.
_____ I am evaluated by my superiors on the basis of how well I get along with others in my own department.
_____ I am evaluated by my superiors on the basis of how well I get along with persons in other departments.

Mode of Conflict Resolution

The question below was used to determine the modes of conflict resolution in each organization. The question was repeated and phrased two ways: first asking about the *ideal* way conflict *should* be handled, and then the *actual* way it *was* handled in the organization. Most of the data presented in Chapters II and V deal with the *actual* data and we shall concern ourselves with them.

The 25 aphorisms were selected following Blake and Mouton (*The Managerial Grid,* Houston: Gulf Publishing Co., 1964) on the *a priori* assumption that there were five modes of resolving conflict: confrontation, compromise, smoothing, forcing, and withdrawal. Five proverbs were selected which seemed to fit each category.

The data were factor analyzed using an orthogonal rotation. Three factors were identified (Table A–10). The other items were not used in the analysis. Factor scores were computed for each organization and these were used for the comparison in Chapters II and V.

TABLE A–10

Factor and aphorism	*Factor loading*
I. Forcing	
Might overcomes right.	.56
The arguments of the strongest always have the most weight.	.47
He who fights and runs away lives to run another day.	.45
If you cannot make a man think as you do, make him do as you think.	.39
II. Smoothing	
Kill your enemies with kindness.	.42
Soft words win hard hearts.	.41
Smooth words make smooth ways.	.41
When one hits you with a stone, hit him with a piece of cotton.	.38
III. Confrontation	
By digging and digging the truth is discovered.	.57
Seek till you find and you'll not lose your labor.	.50
A question must be decided by knowledge and not by numbers, if it is to have a right decision.	.41
Come now and let us reason together.	.41

There is an old proverb that says, "It may be true what some men say; it must be true what all men say." The problem in applying this to the way people work together in organizations is that all men do not say the same thing. Persons in any organization have different ways of dealing with their work associates in other departments. The proverbs listed in the two questions below can be thought of as descriptions of some of the different possibilities of resolving disagreements as they have been stated in literature and in traditional wisdom.

1. You are asked to indicate *how desirable in your opinion* each of of the proverbs listed below is as a way of resolving disagreements between members of different departments. Please use the following scores in evaluating *the desirability of* each proverb.

   (1) Very desirable
   (2) Desirable
   (3) Neither desirable nor undesirable
   (4) Undesirable
   (5) Completely undesirable

Indicate your evaluation in the spaces below:

_____ 1. You scratch my back, I'll scratch yours.
_____ 2. When two quarrel, he who keeps silence first is the most praiseworthy.
_____ 3. Soft words win hard hearts.
_____ 4. A man who will not flee will make his foe flee.
_____ 5. Come now and let us reason together.
_____ 6. It is easier to refrain than to retreat from a quarrel.
_____ 7. Better half a loaf than no bread.
_____ 8. A question must be decided by knowledge and not by numbers if it is to have a right decision.
_____ 9. When one hits you with a stone, hit him with a piece of cotton.
_____10. The arguments of the strongest always have the most weight.
_____11. By digging and digging, the truth is discovered.
_____12. Smooth words make smooth ways.
_____13. If you cannot make a man think as you do, make him do as you think.
_____14. He who fights and runs away lives to run another day.
_____15. A fair exchange brings no quarrel.
_____16. Might overcomes right.
_____17. Tit for tat is fair play.
_____18. Kind words are worth much and cost little.
_____19. Seek till you find, and you'll not lose your labor.
_____20. He loses least in a quarrel who keeps his tongue in cheek.
_____21. Kill your enemies with kindness.
_____22. Try and trust will move mountains.

______23. Put your foot down where you mean to stand.
______24. One gift for another makes good friends.
______25. Don't stir up a hornet's nest.

2. In answering this question you are asked to shift from *what is desirable* to *what actually happens* in your organization. As you read the proverbs below, please indicate, using the following scale, to what extent these proverbs describe behavior in your business.

   (1) Describes very typical behavior which usually occurs.
   (2) Describes typical behavior which occurs frequently.
   (3) Describes behavior which occurs sometimes.
   (4) Describes untypical behavior which seldom occurs.
   (5) Describes behavior which never occurs.

(The list of aphorisms was repeated)

Interview Data

In addition to the questionnaire items described and to the interviews about performance and environmental conditions conducted with top executives, all respondents were interviewed for approximately one hour. The list of questions below developed for the plastics organizations is typical of those covered in these interviews. However, there was often some variation in these questions depending upon the organization, the person's position, or specific problems raised by other respondents in the organization. Our intention in these interviews was to obtain as accurate a description of the functioning of the organization as we could, as well as to learn about the problems faced by organizational members.

1. As a starting point, could you tell me how decisions actually get made on new projects? For example, whom do you talk to about starting and stopping projects; whom do you consult about getting more people or additional funds and equipment; and how frequently are these projects reviewed?
2. In working on these matters how much direction do you receive from your superiors; is this direction sufficient or would you like more direction or less? For example, does your boss give you specific instructions about how to carry out activities or does he leave the details up to you? Do you consider his comments to be instructions which must be followed or as suggestions which may be disregarded?
3. How many persons outside your unit do you come into contact with frequently—say, once a week or more—while you are working on innovations?
4. I would like you to rank the various functions (sales, production,

and research) in terms of their general standing in the company (or division) which units do people in the company (or division) consider to have the highest and the lowest standing?

5. I would now like to spend a little time talking about the coordination of the several units involved in the innovation process. Before we talk about how the various segments work together, I would like to get your thoughts on where close coordination is most important. Could you rank each set of units in terms of the closeness of coordination required to obtain effective action on innovation projects? (List units and work through the ranking with respondent.)
6. Which of these units initiate most of the innovation ideas? What other units are required to follow through on these ideas?
7. In meetings with representatives of different units about new product matters, how frequently do you have disagreements?
8. What is the nature of these disagreements and what positions do the representatives of each unit take?
9. If there are disagreements which are difficult to resolve, do you continue discussions until a solution is reached or do you ask for help from a higher authority?
10. What committees, liaison individuals and other devices are used to improve coordination between your unit and other units? I would also like to get your opinion about the strengths and limitations of these devices.
11. How do all these things add up in terms of your feelings about your job and career?

REFERENCE FOOTNOTES

Chapter I

1 George C. Homans, *The Human Group* (New York: Harcourt, Brace & Co., 1950).

2 Harry C. Triandis, "Notes on the Design of Organizations," in *Approaches to Organizational Design,* James D. Thompson, ed. (Pittsburgh: University of Pittsburgh Press, 1966).

3 Henri Fayol, *Industrial and General Administration,* Part II, Chapter I, "General Principles of Organization"; Chapter II, "Elements of Administration" (Paris: Dunod, 1925).
Luther Gulick, "Notes on the Theory of Organizations," in *Papers on the Science of Administration,* Luther Gulick and Lyndall F. Urwick, eds. (New York: Institute of Public Administration, Columbia University, 1937).
Lyndall F. Urwick, "Organization as a Technical Problem," in *Papers,* op. cit.
James D. Mooney, "The Principles of Organization," in *Papers,* op. cit.

4 Research evidence of this can be found in James G. March and Herbert A. Simon, *Organizations* (New York: John Wiley & Sons, 1958).

5 This difference was suggested by the work of Fiedler. For example, see Fred E. Fiedler, *Technical Report No. 10,* Group Effectiveness Research Laboratory, Department of Psychology, University of Illinois, May 1962.

6 R. H. Hall, "Intraorganizational Structure Variables," *Administrative Science Quarterly,* December 1962, pp. 295–308.

7 This point was first made by Muzafer Sherif in "Superordinate Goals in the Reduction of Intergroup Conflict," *American Journal of Sociology,* March, 1958, pp. 356–394, and *Intergroup Relations and Leadership,* Muzafer Sherif, ed. (New York: John Wiley & Sons, 1962); more recently in the context of industrial organizations by March and Simon, op. cit. It is put forth in an earlier report of part of this study by Jay W. Lorsch, *Product Innovation and Organization* (New York: The Macmillan Company, 1965).

8 Mooney, op. cit., and Urwick, op. cit. One exception was Gulick, op. cit.

9 These same points have been made by Joseph Literer in *The Analysis of Organizations* (New York: John Wiley & Sons, 1965).

10 One exception to this was Mary Parker Follett in *Dynamic Administration: The Collected Papers of Mary Parker Follett* (New York: Harper and Brothers, 1940).

11 A more comprehensive review of this literature and its sources is provided in Chapter VII.

12 An important exception to this statement is A. K. Rice and his colleagues at Tavistock. For example, see A. K. Rice, *The Enterprise and Its Environment* (London: Tavistock Publications, 1963), pp. 186–198. March

and Simon are two other important modern theorists who have recognized the importance of differentiation and integration; see March and Simon, op. cit.

[13] "Marshall Company Case," in *Organizational Behavior and Administration,* Paul R. Lawrence and John A. Seiler, eds. (Homewood: Richard D. Irwin and the Dorsey Press, 1965).

[14] "Pioneer Packing Company," Case No. 449 R, Copyright by the President and Fellows of Harvard College, 1964.

[15] Arthur N. Turner and Paul R. Lawrence, *Industrial Jobs and the Worker* (Boston: Division of Research, Harvard Business School, 1965).

[16] Joan Woodward, *Management and Technology* (London: Her Majesty's Printing Office, 1958); Tom Burns and G. M. Stalker, *The Management of Innovation* (London: Tavistock Publications, 1961). These studies, as well as some other related work, will be examined in more detail in Chapter VIII.

[17] This point was first made by Rice, op. cit., pp. 186–198.

[18] David C. McClelland, *The Achieving Society* (Princeton: D. Van Nostrand Company, 1961).

[19] Robert White, "Ego and Reality in Psychoanalytic Theory," *Psychological Issues,* Vol. III, No. 3, Monograph No. 11, 1963, pp. 24–43.

[20] Ibid., p. 39.

CHAPTER II

[1] The pilot stage of this study is described in detail in Lorsch, *Product Innovation and Organization.*

[2] For a detailed description of the prior research that led to these predictions, see Paul R. Lawrence and Jay W. Lorsch, "Differentiation and Integration in Complex Organizations," *Administrative Science Quarterly,* June 1967, or Lorsch, op. cit.

[3] Fiedler, *Technical Report No. 10,* Group Effectiveness Research Laboratory.

CHAPTER III

[1] For a complete review of this earlier work, see Lawrence and Lorsch, "Differentiation and Integration in Complex Organizations."

[2] The importance of this intermediate position was first discussed in the pilot study for this project. See Lorsch, *Product Innovation and Organization.*

[3] Peter M. Blau and W. Richard Scott, *Formal Organization* (San Francisco: Chandler Publishing Company, 1962), p. 185.

[4] This factor was suggested by Alvin Zander and Donald Wolfe, "Administrative Rewards and Coordination," *Administrative Science Quarterly,* June 1964, pp. 30–69.

[5] Murray Horwitz, "Hostility and Its Management in Classroom Groups," in *Readings in the Social Psychology of Education,* W. W. Charters and

N. L. Gage, eds. (Boston: Allyn and Bacon, 1964, c. 1963), pp. 196–212.

6 This same point has also been made by Claggett G. Smith and Oguz N. Ari, "Organizational Structure and Member Consensus," *American Journal of Sociology,* May 1964, pp. 623–638.

7 This determinant was also suggested by the work of Smith and Ari as well as our own ideas about influence based on competence, which have been mentioned above.

8 These modes of resolving conflict were first classified by Robert R. Blake and Jane S. Mouton in *The Managerial Grid* (Houston: Gulf Publishing Company, 1964).

CHAPTER IV

1 For a full discussion of the relative certainty of different types of technologies, see Joan Woodward, *Industrial Organization: Theory and Practice* (New York: Oxford University Press, 1965).

CHAPTER V

1 This follows the work of Murray Horwitz. See, for example, Horwitz, "Hostility and Its Management in Classroom Groups."

CHAPTER VII

1 Peter M. Blau, *Dynamics of Bureaucracy* (Chicago: University of Chicago Press, 1955).
Alvin W. Goulder, *Patterns of Industrial Bureaucracy* (Glencoe: The Free Press, 1954).
Michael Crozier, *Bureaucratic Phenomena* (Chicago: University of Chicago Press, 1964).

2 Urwick, "Organization as a Technical Problem," in *Papers,* op. cit., p. 51.

3 Ibid., p. 47.

4 J. D. Mooney and Allan C. Reiley, *Onward Industry!* (New York: Harper and Brothers, 1931).

5 Urwick, op. cit., p. 49.

6 Mooney and Reiley, op. cit.

7 Gulick, "Notes on the Theory of Organization," in *Papers,* op. cit.

8 Ibid., p. 41.

9 Ibid., p. 35.

10 Mooney and Reiley, op. cit., p. 31.

11 Ibid., p. 35.

12 Ibid., p. 31.

13 Ibid., p. 36.

14 Ibid.

15 Ibid., p. 37.

16 F. J. Roethlisberger and William J. Dickson, *Management and the Worker* (Cambridge: Harvard University Press, 1939). An account of a

research program conducted at Western Electric Company, Hawthorne Works, Chicago.

[17] Carl Rogers, *Counseling and Psychotherapy* (Boston: Houghton-Mifflin Company, 1942).

[18] *Intergroup Relations and Leadership*, Muzafer Sherif, ed.

[19] Ibid.

[20] Chris Argyris, *Interpersonal Competence and Organizational Effectiveness* (Homewood: Richard D. Irwin and the Dorsey Press, 1962).

[21] Louis B. Barnes, *Organizational Systems and Engineering Groups: A Comparative Study of Two Technical Groups in Industry* (Boston: Division of Research, Harvard Business School, 1960).

[22] Roland Lippitt and Ralph K. White, "Leader Behavior and Member Reaction in Three 'Social Climates'," in *Group Dynamics*, Dorwin Cartwright and Alvin Zander, eds. (Evanston: Row Peterson, 1953).

[23] Max Weber, *The Theory of Social and Economic Organization*, translated by A. M. Henderson and Talcott Parsons (New York: Oxford University Press, 1947).

CHAPTER VIII

[1] *Toward a Unified Theory of Management*, Harold Koontz, ed. (New York: McGraw-Hill Book Company, 1964).

[2] Burns and Stalker, *The Management of Innovation.*

[3] Ibid., pp. 5 and 6.

[4] Woodward, *Management and Technology.*

[5] Ibid.

[6] Ibid., p. 12.

[7] Ibid., p. 10.

[8] Lawrence E. Fouraker, unpublished manuscript.

[9] Ibid., p. 5 ff.

[10] Ibid., Chapter II, pp. 8, 9, and 11.

[11] Ibid., Chapter IV, pp. 1 and 2.

[12] Ibid., Chapter IV, p. 6.

[13] Alfred Chandler, *Strategy and Structure: Chapters in the History of Industrial Enterprise* (Cambridge: The M.I.T. Press, 1962).

[14] Ibid., p. 15.

[15] Ibid., p. 16.

[16] Ibid., p. 41.

[17] Stanley Udy, "Administrative Rationality, Social Setting, and Organizational Development," in *New Perspectives in Organization Research*, W. W. Cooper, H. J. Leavitt, and M. W. Shelley II, eds. (New York: John Wiley & Sons, 1964).

[18] Stanley Udy, *Organization of Work: A Comparative Analysis of Production Among Non-industrial Peoples* (New Haven: HRAF Press, 1959), p. 126.

19 Harold J. Leavitt, "Unhuman Organizations," *Harvard Business Review*, July–August 1962, pp. 90–98.

20 The discussion of this topic has drawn on the analysis set forth by Fouraker, op. cit.

21 Richard E. Walton and Robert B. McKersie, *A Behavioral Theory of Labor Negotiations: An Analysis of a Social Interaction System* (New York: McGraw-Hill Book Company, 1965).

22 Ibid.

23 Fiedler, *Technical Report No. 10*, Group Effectiveness Research Laboratory.

24 Victor H. Vroom, *Some Personality Determinants of the Effects of Participation* (Englewood Cliffs: Prentice-Hall, 1960), p. 60.

25 Turner and Lawrence, *Industrial Jobs and the Worker*.

CHAPTER IX

1 Warren Bennis, *Changing Organizations: Essays on the Development and Evolution of Human Organization* (New York: McGraw-Hill Book Company, 1966).

2 Howard B. Perlmutter, *The Social Architecture of Essential Organizations* (London: Tavistock Publications, 1965).

3 Herbert Spencer, *Autobiography* (New York, 1904), Vol. II, p. 56.

4 Blake and Mouton, *The Managerial Grid*.

5 Abraham Zaleznik and David Moment, *The Dynamics of Interpersonal Behavior* (New York: John Wiley & Sons, 1964).

6 Argyris, *Interpersonal Competence and Organizational Effectiveness*.

7 Robert R. Blake, Jane S. Mouton, Louis B. Barnes, and Larry Greiner, "Breakthrough in Organizational Development," *Harvard Business Review*, November–December 1964, pp. 133–155.

8 Edmund P. Learned, C. Roland Christensen, Kenneth R. Andrews, and William D. Guth, *Business Policy—Text and Cases* (Homewood: Richard D. Irwin and the Dorsey Press, 1964).

9 Paul R. Lawrence, *The Changing of Organizational Behavior Patterns* (Boston: Division of Research, Harvard Business School, 1958).

10 Udy, *Organization of Work*.

11 John W. Gardner, *Self Renewal* (New York: Harper and Row, 1965).

12 Chandler, *Strategy and Structure*.

NAME INDEX

SUBJECT INDEX